T0318803

METALS AND ENERGY FINANCE

Advanced Textbook on the Evaluation of Mineral and Energy Projects

METALS AND ENERGY FINANCE

Advanced Textbook on the Evaluation of Mineral and Energy Projects

Dennis L Buchanan
Imperial College London, UK

ICP
Imperial College Press

Published by

Imperial College Press
57 Shelton Street
Covent Garden
London WC2H 9HE

Distributed by

World Scientific Publishing Co. Pte. Ltd.
5 Toh Tuck Link, Singapore 596224
USA office: 27 Warren Street, Suite 401-402, Hackensack, NJ 07601
UK office: 57 Shelton Street, Covent Garden, London WC2H 9HE

British Library Cataloguing-in-Publication Data
A catalogue record for this book is available from the British Library.

METALS AND ENERGY FINANCE
Advanced Textbook on the Evaluation of Mineral and Energy Projects

ISBN 978-1-78326-850-4
ISBN 978-1-78326-851-1 (pbk)

In-house Editors: Mary Simpson/Chandrima Maitra

Typeset by Stallion Press
Email: enquiries@stallionpress.com

Printed in Singapore

Contents

Acknowledgements

This book incorporates ideas that first evolved during the delivery of a range of professional development courses in London, Vancouver, Toronto, Stockholm and Johannesburg. The most relevant of the courses is probably "Technical & Financial Evaluation of Mineral Projects" (TechFin) that was launched in 1999 as a joint initiative with the Wits Business School Executive Education programme in Johannesburg and Imperial College Centre for Continuing Professional Development. The original TechFin course included basic concepts of discounted cash flow (DCF) analysis with an introduction to finance and accounting provided by Richard Anderson. Workshops were based around a set of custom DCF models for a range of different mineral projects. These had limited functionality and utility and were based on pre-funding scenarios.

An early version of IC-MinEval, a Visual Basic-based tool that automated the generation of financial models, developed with Dr Mark Heyhoe at Imperial College, provided the core intellectual property for an Imperial College start-up company. This was launched in 2000 and we called it IC-FinEval Ltd. I was the Chairman, Mark Heyhoe was Technical Director and Roger Haiat was our Commercial Director with Mervyn Jones joining the Board to represent Imperial College. The development of coal and petroleum fiscal modelling tools followed and functionality was undertaken with the support of Professor Tim Shaw and Colin Howard.

Core concepts outlined in this book arose from the collaboration I enjoyed with all these individuals while developing TechFin and IC-FinEval Ltd and I am most grateful to Mark, Roger and

Mervyn for the generosity they displayed over the years in sharing their invaluable experience with me which helped to shape my ideas. Tim and Colin remain involved in the programme of joint teaching that we deliver and in the development of the concepts outlined in the book. I continue to value this ongoing professional and academic collaboration.

In 2012, IC-FinEval Ltd. was dissolved and the IP transferred to Imperial College but the functionality of IC-MinEval and IC-CoalEval remains available through InfoMine Inc. Simon Houlding, Vice-President of Professional Development for InfoMine and responsible for EduMine, the professional development division, provided outstanding support in this regard. He also supported the delivery of the IC-Mineval-based course "Valuation of Mineral Projects" in Vancouver and Toronto that was launched in 2006.

In 2006 I also launched the MSc Metals and Energy Finance degree as a joint Faculty of Engineering and Business School programme and the structure of this book is based on its teaching modules. Colleagues at Imperial College who have contributed to the delivery of the programme have generously permitted me to use some of their teaching material in preparing this book. Sources are given in figure captions when directly derived from their presentations and these included Professor Jan Cilliers, Professor Olivier Dubrule and Dr Stephen Neethling. Final text on petroleum systems was generated based on extensive discussions with Dr Mike Ala and Dr Peter Fitch during student field excursions along the Jurassic Coast for the MSc students. Dr Fivos Spathopoulos introduced me to the world of unconventional petroleum systems and Professor Mark Sephton reviewed the sections on organic chemistry. My thanks go to all of them for sharing their knowledge with me. I take full responsibility though for the views expressed in this book.

Thanks are due to the companies and organisations that granted me permission to reproduce images either obtained while I was on site at their operations or taken from their own image or photo galleries. These are identified in the Figure captions. Many of the insights described in this book arise from professional exchanges with hosts during site visits. I am also obliged to my son James who checked my thermodynamic and spreadsheet calculations and helped with the minerals engineering perspectives.

All the material in the book has been presented to the 2014/2015 cohort of postgraduate MSc Metals and Energy Finance students who proved to be the most discerning and attentive of recipients. I must acknowledge their active feedback which encouraged me to justify all contentions made and re-visit first principles in a way that would not have been possible had I been working in isolation.

Finally though this book could not have been written without the support and patience of my wife Vaughan and to her my grateful thanks. She acted as my own private copy editor. Her queries about confused sentences inevitably identified sections where there was underlying uncertainty in my own mind about the best approach to use in addressing the topic. Revising the sentence clarified the concept, at least in my own mind, but I hope also for the reader.

List of Abbreviations

AIM	Alternative investment market
AMD	Acid mine drainage
ANFO	Ammonium nitrate and fuel oil
ATBI	After tax before interest
BIF	Banded iron formation
bpd	Barrels per day
BS	Black and Scholes
Capex	Capital costs
CAPM	Capital asset pricing model
CEng	Chartered engineer
CEO	Chief executive officer
CMMI	Council of Mining and Metallurgical Institutes
CRIRSCO	Committee for Mineral Reserves International Reporting Standards
EBITDA	Earnings before interest, taxes, depreciation and amortisation
DCF	Discounted cash flow
dmtu	Dry metric tonne unit
EIA	Environmental impact assessment
EITI	Extractive industries transparency initiative
EPC&M	Engineering, procurement, construction and management
EPT	Excess profit tax
EV	Enterprise value
FCFF	Free cash flow to the firm
FEED	Front end engineering design
FTFS	Full technical feasibility study

FVF	Formation volume factor
F–T	Fischer–Tropsch
GDP	Gross domestic product
GRV	Gross rock volume
ICMM	International Council on Mining and Metals
IFC	Issued for construction
IFC	International Finance Corporation
IFRS	International financial reporting standards
IFT	Issued for tender
IMMa	International Mining and Minerals Association
IOGs	International oil groups
IoM3	Institute of Materials, Minerals and Mining
IPO	Initial public offering
IRR	Internal rate of return
ISA	International Seabed Authority
JORC	Australian Institute of Mining and Metallurgy
JV	Joint venture
LLR	Loan life ratio
LME	London Metal Exchange
LNG	Liquefied natural gas
LSE	London Stock Exchange
MC	Monte Carlo
MMbbls	Million barrels
Mtpa	Million tonnes per annum
N/G	Net-to-gross ratio
NOCs	National oil companies
NPV	Net present value
NSR	Net smelter returns
NYMEX	New York Mercantile Exchange
OpCosts	Operating costs
PAIBT	Profit after interest before tax
PAIT	Profit after interest and tax
PBIT	Profit before interest and tax
Pdf	Probability density function
PEA	Preliminary economic assessment

PERC	Pan European Reserves and Resources Reporting committee
PFS	Preliminary feasibility study
PGE	Platinum group elements
PLR	Project life ratio
ppm	Parts per million
PRMS	Petroleum resource management system
PSC	Production sharing contract
PV	Present value
REE	Rare earth elements
ROM	Run of mine
ROV	Real option valuation
SAAS	Software as a service
SAMREC	South African Institute of Mining and Metallurgy
SCF	Specialist commodity funds
SEC	US Securities and Exchange Commission
SFO	Serious Fraud Office
So	Oil saturation
SPE	Society of Petroleum Engineers
STOIIP	Stock tank oil-initially-in-place
tpa	Tonnes per annum
tpd	Tonnes per day
TSX	Toronto Stock Exchange
VE	Venture exchange
VMS	Volcanogenic massive sulphide
WACC	Weighted average cost of capital
WTI	West Texas intermediate

Chapter 1

Extractive Industry Finance and Mineral Economics

1.1 Introduction

This book covers the field of extractive industry finance, which is a specialist subset of economics. The financial services sectors make little distinction between the mining and petroleum extractive industries, but it should be recognised that mining is just part of the extraction stages that include down-stream processing, while coal is as important a source of energy as petroleum. The choice of *Metals and Energy Finance* for the title of this book therefore allows for a broader treatment of both the technical and financial aspects of the extractive industries.

The concept of "mineral economics" has the connotation of supply and demand although it does incorporate elements of finance. The term "economic geology" is historical in its origin and is concerned mainly with the processes involved in the concentration of metals. The term "mineral deposits studies" is preferred in this book. The term "petroleum economics" tends to cover capital and cost estimates, whereas petroleum fiscal modelling provides the basis for investment decision-making. This book addresses petroleum fiscal modelling.

There are some notable similarities between mineral and petroleum projects. Both are based on depleting resources and therefore companies are under constant pressure to replace reserves. The products are commoditised and globally traded on both spot and futures markets as well as being based on long-term contracts (iron ore, coal and gas supplies). These are products that have low demand elasticity, are capital intensive, require long development lead times and are associated with high technical and commercial risk. There are also environmental compliance and regulatory burdens associated with the development of all natural resource projects.

The book identifies the investment opportunities that are being offered across the whole spectrum of the mining cycle by relating this to the various funding options in the progression from exploration through evaluation, pre-production development, development and, finally, into production. It also addresses the similarities of natural resource projects, whether minerals or petroleum, while identifying their key differences

In the discipline of metals and energy finance it would be expected that practitioners know how to read a set of financial accounts, apply the capital asset pricing model, appreciate the difference between spot price and volatility for commodity prices, and understand derivatives (hedging and option pricing using the Black–Scholes). An understanding of these fields would simply be part of the financial literacy that is the foundation of what is needed in the evaluation of a natural resource project.

1.2 Financial Engineering

Fundamental to the approach used in this book is the contention that financial engineering should be considered equal to conventional engineering, as it has a design component. To ensure optimal investment return in natural resource projects, conventional engineering is blended with financial engineering. Timing is critical in optimising return on investment and innovative use of financial engineering can enhance returns to investors. Financial models can identify where sensitivities on investment performance indicators are a function of technical assumptions used in conventional engineering.

Extractive industry finance recognises that major changes in the international mining and petroleum industries over the last decade have significantly increased the demand for professionals with skills that blend technical and financial engineering with a business perspective. It is now considered perfectly respectable for financial engineering to merit the UK's Engineering Council's Chartered Engineer (CEng) qualification, particularly where finance is combined with the technical aspects of mining, metals and petroleum. A poorly designed financial model can have as big an impact on a business as structural failure in construction.

Financial models of case studies covered in this book demonstrate how financial engineering can be used to offset risk associated with operational issues that impact on the original financial performance indicators. An example would be the reduction in net present value as a consequence of unexpected delays in meeting production targets.

1.3 Role of Finance in Society

After the 2008 international financial crisis and the more recent euro financial crisis, there is a universal perception among young graduates that complex financial products have real relevance to society. It was the failure of investment banks themselves to understand the risk inherent in the quantitative tools they had developed which is now also creating career opportunities for graduates with sound mathematical skills. The 2008 credit crisis has simply emphasised the importance of good training in technical and financial aspects of the commodity business and its role in society.

The fundamental basis for the interest in metals and energy is that the assets are tangible and produce commodity-based products that are an integral part of the natural geological environment. The products are traded internationally and generate cash in hard currency in countries in which the resources are located. However it follows from this that no matter how smart the financial engineering associated with investment structures, if basic estimates of grade, tonnage and recoveries for metals or the volumetrics and recovery assumptions for hydrocarbons are wrong because of a flawed understanding of mineral or petroleum geosciences, then the whole basis for a valuation and investment decision will also be flawed.

Society, and certainly ambitious young graduates, understands that it is finance that is the critical factor in modern economies. There is a high demand for graduates with metals and energy finance skills from the investment banking community, the new business development divisions of major mining and petroleum groups, and independent business analysis and consultancy groups focused on the mining, metals, power, energy, cables, fertiliser and chemical sectors.

1.4 Minerals

The mining industry is at a watershed, with focus on growth since 2006 altering in 2013 to optimisation of existing investment. Before the financial crisis of 2008, growth was achieved mainly through mergers and acquisitions. Valuations placed on the assets before the financial crisis were demonstrated to have been manifestly inflated. This is reflected in the massive impairment write-downs that began in 2009 and continue into the present. This has coincided with some well publicised changes at the chief executive officer (CEO) level of the major mining companies, with the new incumbents often bringing operational experience to the appointment.

While in the past the share prices of companies in the minerals business have benefited from the high commodity prices, it is not that clear if value has been added to the underlying merits of the assets. For this to happen, new projects based on previously undeveloped resources will need to be brought onstream or additional ore extracted from existing assets. This organic growth creates real value for the company as the primary asset base of mineral reserves is enhanced through investment. This can be at a significantly lower cost to a similar quality asset where this is obtained through an acquisition.

The revised business strategy appears to be that of consolidation within existing operations. The major mining companies know they can do little about market demand, but they can enhance shareholder value by focusing on cost management (both operating and capital). This is epitomised by the approach of some majors spinning off a suite of what they consider non-core assets, leaving them to focus on core assets. Operational priorities should result in efficiencies. Ramping up production of bulk commodities such as iron ore and coal to achieve economies of scale in an over-supplied market in order to put pressure on smaller higher cost producers is an interesting business strategy being followed by some of the major producers.

Gold mines are in a class of their own — high-grade deposits with metallurgically simple ores are normally good investments regardless of supply and demand considerations.

The key driver for many major base metal operations is the transition from open pit to underground mining based on block and panel caving. Chilean copper mines are achieving higher underground production rates by utilising new technologies and better knowledge of cave behaviour.

Value release for industrial minerals such as iron ore are completely dependent on infrastructure such as dedicated rail links and deep water harbours. This will almost certainly require the involvement of international organisations like the World Bank for the development of projects such as the various West African iron ore deposits. The same applies to coal deposits such as those found in Mongolia. Release of value arises from building power stations, an electrical grid and rail connections that are not core businesses for a mining company. In South Africa more smelting capacity is needed but this requires further investment in electrical power generation, but the private sector cannot compete with Eskom, the public-sector utility, on cost. Nickel concentrate from the Nkomati mine in South Africa has, for example, been shipped outside the country for smelting rather than being done locally.

Any approach to the development of mining projects needs to be clearly linked to reconciling the meeting of international demand for metals with the understandable concern by countries hosting the primary deposits that these are non-renewable resources and, therefore, wasting assets. Chile is a good case history. No matter how large the deposits located in that country, they have a finite size, so how does this generation ensure that value from current exploitation is preserved for future generations? A Norwegian-type petroleum fund or diversification of the economy are just some of the alternative possibilities to leaving the copper in the safest possible place — in the ground in the conditions that nature first created the concentration.

Copper minerals will not deteriorate with time or the metal leak away if left in this state.

Certainly any permitting must be linked to minimising environmental impact during extraction and how industry can fully rehabilitate the vast open pits and waste dumps that remain after closure. Is it then practical to have site monitoring in perpetuity? We do not have to provide definitive answers, but these are legitimate issues the next generation of natural resource professionals are going to have to face.

1.5 Hydrocarbons

The decision by mining companies to ramp up production of bulk commodities seems to mimic the same strategy as some of the national oil companies (NOCs) ramping up production in 2015, notwithstanding downward oil price movements. Petroleum exploration budgets were reviewed and reduced in 2015 and the focus for producers will be on enhanced production methods.

Lower oil prices provide opportunities for mergers and acquisitions that can create scope for rationalisation and cost-cutting. This gives companies access to assets that they would never have been able to develop themselves through pure exploration. For large international oil groups (IOGs), deals can be particularly attractive, offering a way to increase production that can be cheaper and more predictable than high-cost, high-risk exploration.

The volatility in the oil price creates uncertainty over valuations. Majors with strong balance sheets will look to acquire assets but problems of agreeing valuations can stymie deal-making. As the price settles, financial strains on many companies, particularly those exposed to large capital expenditure but with limited cash flows, are likely to provide the driver for takeover activity.

Valuation of assets for acquisition (including the US shale oil industry and unconventional petroleum projects) will become a business priority for IOGs, who will be identifying potential targets and will be using their financial strength to pick up assets at attractive

Table 1.1. Consumption by fuel type in 2013. Source: BP Statistical Review of World Energy, 2014. http://www.bp.com

Commodity	Million tonnes oil equivalent	
Oil	4,185.1	33%
Natural gas	3,020.4	24%
Coal	3,826.7	30%
Nuclear energy	563.2	4%
Hydro electricity	855.8	7%
Renewables	279.3	2%
	12,730.5	100%

prices. Low oil prices therefore represent an opportunity for some parts of the petroleum sector.

The emphasis of this book is not only on metals but also on energy, and is not confined to conventional oil and gas associated with petroleum systems. Part of the book is devoted to coal, which in 2013 supplied 30% of world energy needs (see Table 1.1) and is predicted to grow. As will be noted oil, natural gas and coal, which are collectively described as fossil fuels, provide 87% of energy needs.

Upstream extraction techniques for petroleum and coal are normally different, with the former recovered from wells while the latter is associated with mining, but the Athabasca tar sands operations in Canada are essentially mining operations although classified as petroleum projects.

There is also a convergence between petroleum and coal in the downstream processing stage of Sasol's Secunda plant in South Africa where the Fischer–Tropsch (F–T) process is used to convert coal to petroleum products. This includes the development of conventional petroleum gas resources to enhance the gas-to-liquid capabilities of the F–T process. Theoretically, carbon dioxide generated by the process could be cooled and liquefied and then utilised to recover residual hydrocarbons from mature oil reservoirs.

Pilot projects are being linked to coal mining, carbon sequestration and methane production. Carbon capture and storage, and the utilisation of methane emissions permits operators to attract carbon credits and participate in active carbon trading.

For these reasons the investment community often makes little distinction between petroleum and coal mining, as both sectors are energy-related and commodity-based. The book covers the detailed financial modelling of coal projects and an in-depth treatment of hydrocarbon fiscal regimes based on both tax and royalty and production sharing contracts. The interplay between gas, including liquefied natural gas (LNG) prices and oil prices only adds to the mix.

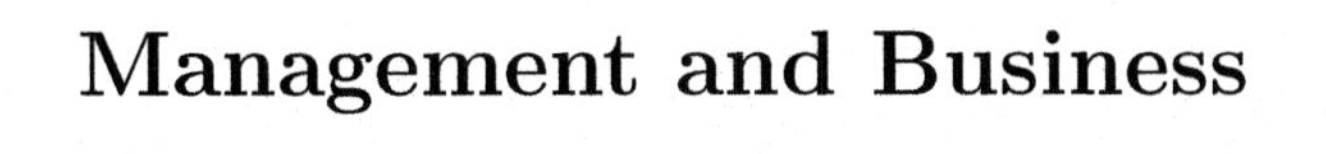

Management and Business

Chapter 2

Cash Flow Modelling and Financial Accounting

2.1 Introduction

This chapter considers the links between the "accounting model" (potentially net profit, earnings per share, price to earnings ratio, even return on investment %) and the "economic model" (present value of the project's expected future cash flow). It outlines the way in which cash flow information is constructed from basic project details and then to represent that information as a set of accounts, as this is widely used for presentation and reporting of financial information. It could be argued, however, that cash flow information is much more useful.

For mineral projects particular attention has to be paid to the treatment of the key independent variables such as grade, and dependent variables, such as grade-tonnage relationships, and the way these influence the rate of mining, associated costs and optimisation of the value of a project. Petroleum projects are based on initial volumetrics, segment production and annual production profiles, revenue, operating costs and capital costs.

The distinction between technical appraisal and financial factors will also be addressed and the reason why discounted cash flow (DCF) models need to be integrated correctly into financial accounts explained. This will be linked to concepts of shareholder value and the role of gearing to maintain an efficient balance sheet.

2.2 Principles of DCF Modelling

As a starting point in building a DCF model for a mineral project it is often assumed that the projects are at the pre-production and pre-investment stages. The principle of discounting cash flows is based on the logic that money received in the future is worth less than that same amount received today, due to the opportunity of earning additional revenue on that sum if it were to be invested elsewhere. Suppose there is a choice of receiving $1,000 today and investing it or receiving $2,000 in 10 years' time. Which is the most valuable outcome? The answer clearly depends on the prevailing interest rate. If it happens to be 5%, the money would be worth $1,629 at the end of 10 years and so it would be better to wait. On the other hand, if the current rate happens to be 10% the sum would be worth $2,594 in 10 years' time and so it would be preferable to take the money now and invest it. The break-even interest rate in this scenario is about 7.2%.

Modelling incremental DCF analyses the financial viability of a project by not only testing that generated revenues are substantially greater than costs and debt service requirements, but also by measuring the present value of those profits. The underlying philosophy in DCF analysis is that the project is to be compared with investing the same stream of cash flows elsewhere.

One of the essential questions in DCF analysis is how to choose the discount rate. At the pre-funding stage this is essentially set by the investor or the new business division of the natural resource company. DCF can then be used to determine the net present value (NPV) of the project, which is essentially a present valuation of the potential of the deposit to generate future profits. NPV is calculated as follows:

$$NPV = C_0 + \frac{C_1}{(1+i)} + \frac{C_2}{(1+i)^2} + \cdots + \frac{C_n}{(1+i)^n},$$

where C is the net cash flow at year n and i is the project discount rate.

Projects with an NPV greater than zero will produce greater revenues than their costs at the minimum acceptable rate of return (the discount or hurdle rate). Mutually exclusive investment opportunities are ranked by magnitude of NPV.

The internal rate of return (IRR) and payback period of a project can also be calculated from a model of future cash flows. IRR is essentially the discount rate at which NPV at time zero of all cash flows is equal to zero, and is calculated as follows:

$$0 = C_0 + \frac{C_1}{(1+i)} + \frac{C_2}{(1+i)^2} + \cdots + \frac{C_n}{(1+i)^n}.$$

Solving for i gives the project's IRR. A project is profitable if the IRR exceeds the opportunity cost of capital (the project's discount rate), and mutually exclusive scenarios are ranked by magnitude of IRR.

Payback period is simply the time taken for the initial capital investment to be recovered by the stream of annual positive cash flows, and is not generally used alone for making an investment decision, as it takes no account of the time value of money.

Clearly the choice of discount rate is a fundamental factor in the evaluation of an investment opportunity whether a mineral or a petroleum project. The impact on the present value factors used to discount a net cash flow for the selected period is demonstrated in Figure 2.1.

Some key points arise from these curves. For a discount rate much above 6% the value of net revenue generated after 15 years has diminishing impact on the present value of the project. It could be argued that for long-life projects of 25 years and greater the technique will significantly undervalue the opportunity. This effect is amplified as the discount rate is increased to the point that the curves start becoming asymptotic. Applying a discount rate much above 15% requires an exceptional project before the NPV justifies making an investment.

There is a separate insight that is provided by the curves at the start of the project life when they are steep. Unless the scenario is modelling refurbishment of an operation already in production with an existing cash flow there will be a period during development and construction when no revenue is being generated. The delay will have

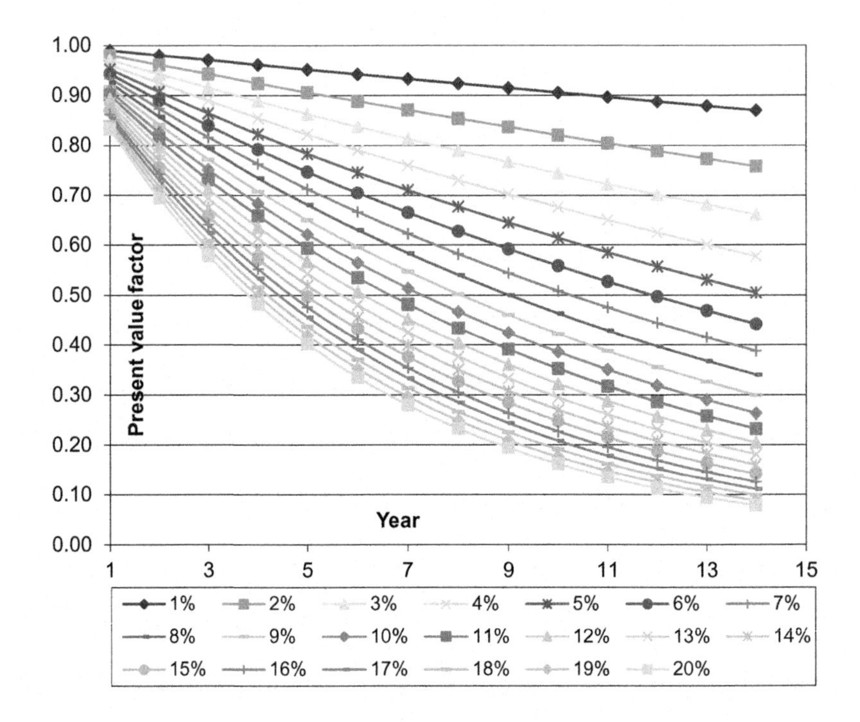

Figure 2.1. Impact of different discount rates over time.

been estimated from the engineering design and any delay in the original assumptions will have a profound impact on the NPV. Reasons include incorrect assumptions that ore characteristics have on plant design and the distortions that arise when clients start requesting changes after detail work has started. Delays, once expenditure has been committed that defers revenue generation, can compromise the whole financial integrity of a project that has to be reviewed in the context of the original risk factors.

The IRR can have multiple roots. Where a model has a positive NPV, with an IRR that exceeds the discount rate, then as the discount rate increases and approaches the IRR the NPV will converge on zero. Continuing to increase the discount rate beyond this point will generate a negative NPV but in some circumstances the NPV can start reversing this trend and the curve can revert and again approach zero. This might be because the capital cost assumptions used for the initial periods of the project became steeply discounted. The NPV of projects with high closure costs may, counter-intuitively, improve as the discount rate increases as the impact is reduced once the present value factors 10–15 years into the future have been applied.

In a marginal project the cumulative cash flows may never become cash positive so the point of pay back is never reached. As a consequence, the model is relatively insensitive to the choice of discount rate. A promoter in these circumstances may elect to use a discount rate of zero, implying that the project remains worthy of investment as it is insensitive to choice of discount rate, but this approach is somewhat disingenuous. In a good project with an early pay back and significant positive cash flows from that point a lower discount rate would significantly enhance NPV.

Unlike most petroleum projects, a mineral project may have a relatively slow ramp-up period before production gets up to "cruising altitude". Failure to achieve the projected ramp-up after the start of production resulting in a delay in reaching design capacity, will further compromise NPV owing to the shape of the present value curves in the early stages of a project. The corollary of this is that by exceeding ramp-up expectation additional value can be created.

Overall performance of a project is often best assessed from a graphical presentation of the cash flows as depicted in Figure 2.2. Initial negative cash flows associated with capital expenditure are replaced by positive cash revenues as production starts. Undiscounted net cash flows depicted in Figure 2.2 will contribute to cumulative net cash flows which establish the breakeven point in a project. Both this milestone and the maximum cash exposure have important practical implications for investors.

The model shown is unlikely to have a positive NPV at a 10% discount rate and may never break even. It is important to note that this is an economic model not an accounting model. In accounting terms Figure 2.2 is essentially unable to be implemented. Capital is needed to fund the cost of development. The overall cash flows should never go negative in a financial model designed to make a final investment decision. The NPV is, however, determined from the economic part of the financial model. To determine the NPV from cash flows from

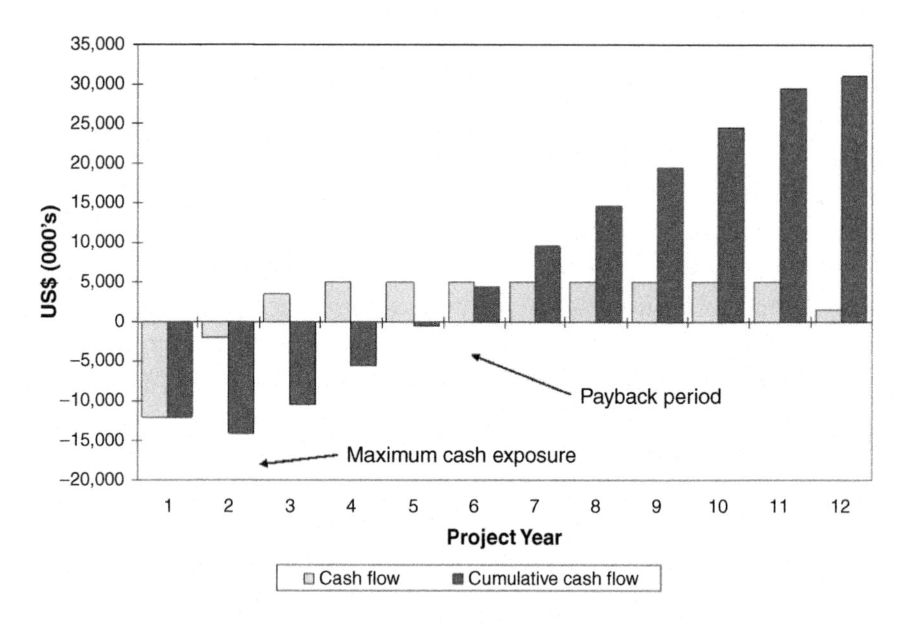

Figure 2.2. Cash flow for a mineral project.

the post-funding scenario would generate an infinite IRR as there is no value that would satisfy the condition needed to generate a zero NPV. There are some circumstances where the financial model of an operation in production where refurbishment is paid for from cash flow never goes cash negative. These can generate an infinite IRR and are good projects to acquire. As a performance indicator IRR is obviously not particularly useful in such circumstances and could be quite misleading.

Even where no artefacts are generated, such as an infinitely high IRR, DCF performance indicators will not in themselves provide an objective basis for an investment decision. Three hypothetical scenarios are given below.

Project A may be attractive to one of the major international mining companies as the potential NPV would be material but not essential for their business. Project B may be too small for the international mining company, but attractive to a mid-size company with existing operating mines and market capitalisation of, say, $1 billion, as the NPV and reasonable IRR could make a significant difference to

Table 2.1. Scenarios illustrating the range of economic performance indicators for three different mining projects.

Project	A	B	C
Capital cost (million)	$500	$250	$25
Mine life (years)	25	15	10
Discount rate	8	10	15
NPV	$1.5 billion	$750 million	$75 million
IRR(%)	12	15	50

their share price. For an exploration company listed on the London Stock Exchange's (LSE) Alternative Investment Market aiming to bring their first mine into production Project B with its high IRR would be attractive. The NPV is, however, too low to be of interest to the mid-sized company and simply a distraction for the international group.

A financial model is clearly no substitute for understanding the business case for investing in a natural resource project. It does, however, provide a good reality check on the level of return that can be expected. For this reason it should be kept as simple as possible at the early stages of project evaluation.

When generating a DCF financial model a distinction should be made between the use of "nominal" versus "real" value, "inflation or deflation" in considering costs and "escalation" when dealing with metal prices. Furthermore, regardless of the location of the project, it is often best to use US Dollar capital costs, operating costs and obviously metal prices. As a first pass for countries with high rates of inflation (such as South Africa and Russia) it permits the removal of the impact of variable exchange rates between the local currency and the US Dollar.

If "real" value is defined as the nominal figures adjusted for inflation, then the implementation of tax is going to require deflating net cash flows back to nominal before the NPV can be determined. This makes it more difficult to retain an intuitive feel for the model. It is not therefore recommended for the early stages of project evaluation.

When building into the model the role of derivatives, a hedging strategy can be implemented. This becomes relevant where project finance is involved and allows consideration of forward curves for currency and metal prices which will be either in backwardation (assumes that the prices will be lower in the future compared to today's spot price) or going into cantango. That fixes in the metal prices based on the proportion of metal hedged and the length of the hedge. There might even be a currency hedge.
In summary:

- Real = nominal adjusted for inflation
- Inflation and deflation: applies to costs
- At the pre-feasibility stage use nominal when generating a pre-investment DCF model
- Escalation: applies to metal prices, and
- At the investment stage may hedge metal.

2.3 Valuation of Mineral Projects Based on Technical and Financial Modelling

2.3.1 *Introduction*

The valuation of mineral projects is best undertaken based on a DCF model. This requires a deposit with an indicated resource as defined by the reporting codes (see Section 3.4.1). Alternative methods include comparable transactions which normally apply to exploration properties and require identification of market capitalisation of several junior companies with licences covering similar metallogenic characteristics. Finally there is the appraised value method which assumes that the value of a property is either enhanced or diminished by an exploration programme. Funds expended that develop the potential of the project will enhance value proportionately. The reality is that in most cases the exploration process will eliminate targets and that expenditure has to be written off.

The NPV generated by a DCF model equates to the intrinsic value of the project. This is normally used as a base line for determining the "market value" of the asset and for a single-project company that should set the price of the shares if listed. Any market turbulence increases the need to understand the interrelationship between technical and financial risk. The key consideration is the intrinsic value of a mineral project. Where current market conditions undervalue the true long-term worth of assets, which are then reflected in low share prices, strategic planning can identify opportunities.

In a situation where a buyer is negotiating with a vendor and a value agreed, this acquisition cost needs to be considered in relation to the NPV. If this is the same then the return on the investment will be the discount rate selected. There may well be resistance from the buyer to paying the full market value. If, however, the buyer is working to a separate strategic agenda then they may well be prepared to pay more than the NPV.

2.3.2 *Treatment of sunk capital*

The scenario present for an operating mine is somewhat different from a normal mineral project valuation as it is a going concern with sunk capital. The implication of the role of sunk capital is that future production will be put into the slipstream of the existing mine infrastructure. A valuation will need to provide very clear indications on the weight (if any) placed on the sunk capital.

If the valuation is going to be determined simply from the future net cash flows projected from the present, then the arithmetic of the model will ensure it is going to generate a very much higher number than if the original capital cost of establishing the project is offset against the NPV. It needs to be recognised in preparing the valuation, however, that in a takeover or acquisition situation, sunk capital could be added to the valuation by the vendor, as a buyer should be prepared to accept that a mine operating as a going concern has much more value than an undeveloped pristine *in situ* resource. There is no right or wrong outcome and, as a consequence, disputes may arise when a joint venture partner wishes to disinvest which can end up having to be resolved in a formal legal arbitration process.

The need for a valuation of a going concern arises when the company feels that the market has discounted the share price and it needs to reassure shareholders. This becomes critical if the operation has exhausted its working capital and could go into receivership. Closure probably needs to be considered if the company is in this type of financial distress and the cost associated with placing the mine on care and maintenance would need to be deducted from the valuation. The lower valuation could provide a "book-end" to the exercise.

There is some debate about the use of DCF for valuing very large deposits because it tends to undervalue long life projects. A valuation should therefore, if possible, be considered in relation to comparable transactions to determine that it is consistent with market perspectives. The valuation would also be considered in relation to the approach used in asset acquisition within the mining industry. It is recognised that for strategic reasons a company may choose to pay more to acquire a mineral or mining asset than the value generated by a DCF financial model.

In the aftermath of the 2008 financial crisis, projects under construction that had not made sufficient provision for the capital cost found that they were unable to raise additional capital from either the banks that provided the original debt or from existing shareholders. The consequence in some cases was liquidation of the holding company but of course in others an opportunity for any natural resource company with large cash holdings. Given the combination of a distressed sale and the promise of relatively minor levels of investment to complete construction of a project, the acquisition cost may be well below the theoretical valuation based on a DCF model.

In summary, the valuation of an existing project would assume 100% ownership by the holding company, and the NPV compared with the current market capitalisation. This should provide the basis for determining if the shares are undervalued or overvalued compared to the underlying models. Debt can be ignored for this part of the exercise. In particular there would need to be an explanation for the original capital cost in building the operation. It could be treated as sunk and therefore the NPV is only dependent on future cash flows. Provided there is a reasonable mine life remaining, this should generate enhanced NPVs. The market may not, however, agree with this perspective (which is where the amount of debt becomes relevant) and this would be reflected in a share price lower than a model might predict.

2.4 Analysis of Risk and Uncertainty

The minimum level of information needed to generate a basic DCF model for a small hypothetical underground gold mine is given in Table 2.2.

It should be noted that in attempting to undertake an independent evaluation of a project, establishing even this level of information can be a real challenge.

Once the financial model has generated the NPV and IRR these performance indicators are presented as deterministic outcomes often to an unrealistic level of precision. A much more useful approach is to undertake sensitivity analysis on key variables. The common approach is to generate a sensitivity diagram in which the selected variable is flexed while the associated variables remain fixed. To ensure that the impact of the selected variable remains comparable to the others in the analysis the project life must remain the same. Flexing dilution, the mining rate, ramp-up and the size of the resource will change the project life. This essentially constrains the use of a sensitivity diagram to grade, plant recovery, operating cost and capital cost. As the key row in the financial model is revenue derived from grade multiplied by plant recovery multiplied by metal price flexing, any one of these variables by the same relative proportion will have the same impact on revenue and therefore on NPV.

The results of sensitivity analysis on grade, capital cost and operating cost are given in Figures 2.3 and 2.4.

The base case is clearly marginal. A characteristic of this type of scenario is that a 20% increase in the capital cost of the project will halve an already poor NPV. If grade proves to have been overestimated by 10% the project will not be viable.

Taking the same base case and using a gold price of $1,000 per ounce generates the sensitivity diagram given in Figure 2.4. As can be seen, a 20% increase in the capital cost has a trivial impact on NPV. If grade proves to have been overestimated by 10% the project remains robust. It is these macro factors that should be considered

Table 2.2. Minimum information required for a base case financial model of a gold project.

Variable	Value and units	Comment
Mining method		Underground.
Construction period	2 years	
Resource	3.24 million tonnes	This is the amount of *in situ* mineralised material.
Mining dilution	9%	**9%** which means that for every 9.1 tonnes of ore we extract we will also extract 0.9 tonnes of sub-economic rock. Having some dilution is normal, as it is often caused by mining conditions, but the key is to keep it to a minimum as it is also often caused by poor grade control.
Grade	10 grams/tonne	In the calculations a conversion 31.1 grams per troy ounce is used.
Plant recovery	90%	The *in situ* grade here is **10 g/t** and we have been given a plant recovery of **90%**. Based on these figures, we have 32.4 tonnes of gold *in situ* of which we are able to extract 29.16 tonnes.
Ore mined	1,000 tonnes/day	Assumes ore mined at a rate of 360,000 tonnes per annum and, unrealistically, no ramp-up period.
Mine life		Based on 9% dilution over the entire life of mine we will be extracting 0.29 mt of sub-economic material, or a combined total or 3.53 mt. This gives a mine life of around 10 years.
Capital cost	$55 million	Estimate for the project (mine, mill and processing plant) and we assume this is split **40%:60%** over the 2-year construction period.
Operating cost	$44 per tonne	Estimate for the cost of mining and processing.
Working capital	25%	This is set as a proportion of annual operating costs equivalent to 3 months. It is assigned at the start of production in year three of the project and then returned at the end.

(*Continued*)

Table 2.2. (*Continued*)

Variable	Value and units	Comment
Metal price	$300 and $1,000 per troy ounce	Base case and escalated scenario.
Royalty rate	2%	Determined from net revenue and essentially a resource rent.
Tax rate	35%	This is based on the operating margin.
Depreciation		This is based on straight-line depreciation which allows us to use the Capex to reduce the taxable income by the same amount each year of production.
Discount rate	8%	

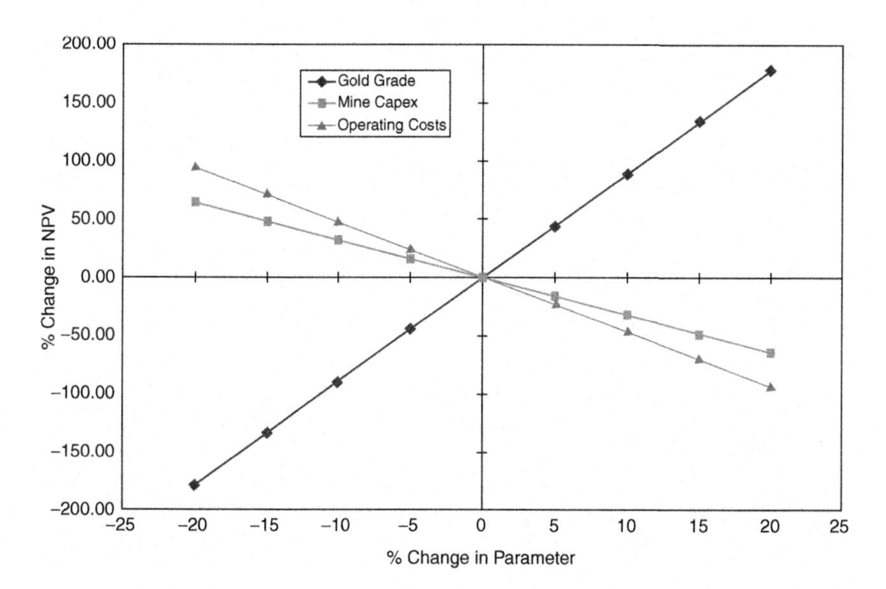

Figure 2.3. Sensitivity diagram for the base case of a hypothetic gold project which at $300 per ounce generates an NPV of $11.61 million and an IRR of 12%.

in undertaking an initial evaluation of a mineral project. A marginal project is seldom enhanced to any significant degree by focusing on reducing capital cost estimates. It is more productive to consider the

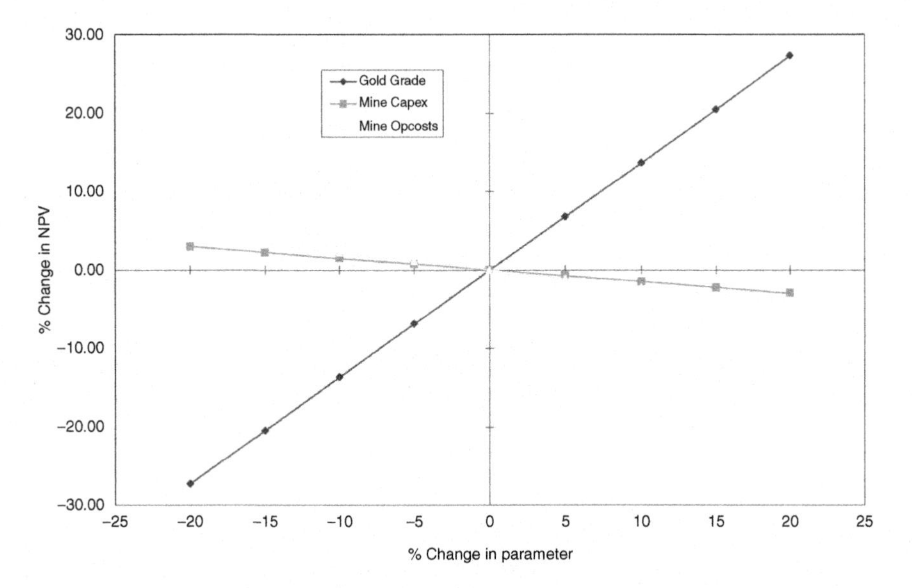

Figure 2.4. Sensitivity diagram for the base case of a hypothetic gold project that when run assuming a gold price of \$1,000 per ounce, generates an NPV of \$263 million and an IRR of 72%.

potential of the deposit to host a larger resource of mineralisation present at a higher grade than the scenario being modelled.

The use of a sensitivity diagram does recognise the merits of moving from a deterministic to a probabilistic approach, but being constrained to flexing only one variable at a time is clearly not realistic. In the real world grade, plant recovery, operating cost and capital cost will be associated with their own level of uncertainty. Monte Carlo (MC) simulation provides a powerful technique to assess risk and uncertainty by recognising that, as a first approximation, each variable will contribute to a range of NPV outcomes. A good example of the application of the technique is given by Hayes (2014) who explains it as follows:

"The inputs of a DCF are forward looking estimates, and subject to uncertainty. For example, capital and operating costs are estimated prior to constructing a mining project, but are not known with certainty until they are fully incurred. Monte Carlo Stochastic simulation attempts to account for this uncertainty in

DCF inputs by enabling use of probabilistic inputs and producing probabilistic outputs. Important inputs are identified and assigned as random variables with distributions as defined by the modeller. These distributions can be based on historical trends, forward projections, industry standards and professional judgement. To produce results, a set number of iterations of the DCF are calculated. The value of the stochastic input variables changes with each iteration, varying within the bounds of the distributions assigned by the modeller. Consequently, the NPV, and any other output variable produced by each iteration of the DCF is different. The values of the output variable(s) are saved and collected until an adequate number of iterations is completed."

An example of the type of output generated from a MC simulation is given in Figure 2.5. This recognises that a range of outcomes

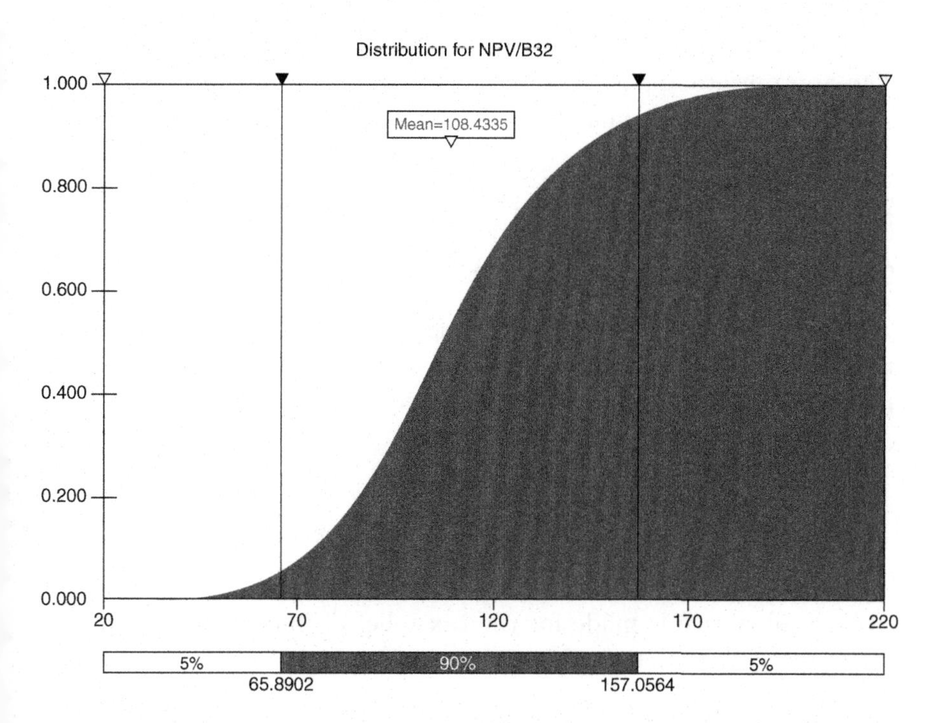

Figure 2.5. Distribution of NPV based on a MC simulation of a gold project in which grade, plant recovery, operating costs and capital costs are treated as independent variables.

is possible based on the uncertainty associated with the estimation of grade, plant recovery, operating cost and capital cost which are treated as independent variables. The mean value of NPV should converge on the base case generated in the deterministic model but recognises that low grade may coincide with low plant recovery. In the real world there may be a correlation between plant recovery and its associated operating and capital costs. The method permits modelling these as dependent variables.

This technique comes with a health warning — not all investors welcome having the uncertainty associated with a natural resource project being demonstrated in this way, particularly if one of the outcomes is a finite probability of a negative NPV. On the other hand it can also demonstrate the robustness of a project and the probability of a high NPV.

There is seldom any merit in using sensitivity analysis or MC simulation to flex metal prices as the outcomes are simply self-fulfilling. The appropriate application of MC simulation to metal prices is to set up a real option valuation as outlined in Sections 4.8 and 10.3.

2.5 Project Finance and the Cost of Equity

2.5.1 *Discount rate and cost of capital*

There are two methods of discounting that can be used to calculate the NPV in a financial model. The pre-determined discount rate can be used or the weighted average cost of capital (WACC). This is calculated as follows:

$$\begin{aligned}&(\text{Tax-adjusted cost of debt} \times \text{percentage debt})\\&\quad + (\text{cost of equity} \times \text{percentage equity}).\end{aligned}$$

As the NPV is calculated on the cash flows before funding but after tax, an allowance is made for the tax relief of interest payments on debt. The cost of debt is therefore calculated as:

$$\text{Tax-adjusted cost of debt} = \text{interest rate} \times (1 - \text{tax rate}).$$

The WACC thus varies according to the debt to equity ratio of the project's funding structure. The cost of equity is generally higher than the cost of debt, reflecting the higher rate of return required by the equity holders in comparison to the "cheaper" interest rate on debt. Thus the greater the percentage of total capital funded by debt, the lower the WACC and thus the more favourable the calculated NPV. This is an essential principle behind the consideration of the use of project finance.

Treatment of tax requires the integration of an economic model with an accounting model. At the investment stage the cash flows should not go negative and where debt is used in combination with equity then the cash flows must be integrated with financial accounts. In Table 2.3 simple links from the cash flows through to profit & loss account and the balance sheet are shown. The model includes project funding showing the cash flow impact of raising equity and debt to fund the development of the project and their presentation in the balance sheet. The model also includes the effect of depreciation and tax on profitability together with the treatment of working capital, inventory build-up (equivalent to an ore stockpile), creditors (also called payables) and debtors (also called receivables).

By integrating cash flow with financial accounts taxable income can be determined correctly on the current year's profits. This in turn permits the correct treatment in the financial model of profits after tax before interest (ATBI), profit before interest and tax (PBIT), profit after interest before tax (PAIBT) and profit after interest and tax (PAIT). Where debt is used for funding and the cash flows are discounted by WACC it is clearly important that this is based on ATBI in order to determine NPV. Using WACC to DCFs that have already derived benefit from tax relief on interest would generate an artificially enhanced (and incorrect) NPV.

A financial model which integrates cash flow with financial accounts also permits a more realistic treatment of depreciation from straight line over the life of the project to loss carried forward where the taxable income is equal to the current year's profits (PBIT or PAIBT as appropriate) plus the previous year's losses (with losses carried forward indefinitely). This will generate an enhanced NPV, as the tax benefit is obtained earlier in the life of the project when present value factors give greater weighting to profits ATBI.

Table 2.3. Integrated economic and accounting model for a simple gold project demonstrating the interrelationship between DCFs and the financial accounts. Reproduced with permission of Richard Anderson.

Year	0	1	2	3	4	5
Annual mined ore (t)	0	50,000	100,000	150,000	150,000	150,000
Annual depletion	0	3.7%	7.4%	11.1%	11.1%	11.1%
Total mined, including waste (t)	0	75,000	150,000	225,000	225,000	225,000
Recoverable gold (000's g)	0	248	496	744	744	744
CASH FLOW (000's US$)						
Capital expenditure	−18,300	−550	0	0	0	0
Cash from sale of gold		1,744	4,419	7,210	8,257	8,373
Operating expenses paid:						
per tonne mined		−503	−1,028	−1,553	−1,575	−1,575
per tonne treated		−611	−1,248	−1,886	−1,913	−1,913
Fixed Overhead		−250	−250	−250	−250	−250
Tax paid		0	0	0	−64	−402
Cash flow before funding 1	**− 18,300**	**−170**	**1,893**	**3,521**	**4,455**	**4,233**
Debt funding	14,999	0	0	0	0	−5,000
Interest charge	−1,200	−1,200	−1,200	−1,200	−1,200	−1,200
Cash Flow before Equity 1	**−4,500**	**−1,370**	**693**	**2,321**	**3,255**	**−1,967**
Equity funding	6,001	0	0	0	0	0
Annual Cash Flow	**1,500**	**−1,370**	**693**	**2,321**	**3,255**	**−1,967**
Accumulated cash	1,500	131	823	3,144	6,399	4,433
NPV Factors @ 8.00%	**1.0000**	**0.9259**	**0.8573**	**0.7938**	**0.7350**	**0.6806**
Annual PV's	**−4,500**	**−1,268**	**594**	**1,843**	**2,392**	**−1,338**
Acccum NPV	**−4,500**	**−5,769**	**−5,175**	**−3,332**	**−940**	**−2,278**

(*Continued*)

Table 2.3. *(Continued)*

Year	0	1	2	3	4	5
PROFIT & LOSS ACCOUNT						
Sales revenue	0	2,093	4,884	7,675	8,373	8,373
Less Cost Of Goods Sold	0	−1,059	−2,284	−3,447	−3,738	−3,738
Less depreciation (not a CashFlow!)	0	−698	−1,396	−2,094	−2,094	−2,094
Less interest (is in the Discount Rate!)	−1,200	−1,200	−1,200	−1,200	−1,200	−1,200
Prof before tax	**−1,200**	**−864**	**4**	**934**	**1,341**	**1,341**
Assessed loss cfw	**−1,200**	**−2,064**	**−2,061**	**−1,127**	**214**	**1,341**
Taxable income	**0**	**0**	**0**	**0**	**214**	**1,341**
Tax payable	**0**	**0**	**0**	**0**	**−64**	**−402**
Less Prov for Tax	0	0	0	0	−64	−402
Profit after tax	**−1,200**	**−864**	**4**	**934**	**1,277**	**939**
BALANCE SHEET						
Assets:						
Cash balance	1,500	131	823	3,144	6,399	4,433
Debtors	0	349	814	1,279	1,396	1,396
Inventory	0	353	644	934	934	934
Fixed assets (net)	18,300	18,152	16,756	14,661	12,567	10,472
Total Assets	**19,800**	**18,984**	**19,037**	**20,019**	**21,296**	**17,235**
Liabilities:						
Creditors	0	48	97	145	145	145
Debt	14,999	14,999	14,999	14,999	14,999	10,000
Equity	6,001	6,001	6,001	6,001	6,001	6,001
Accumulated P&L	−1,200	−2,064	−2,061	−1,127	150	1,089
Total Claims on Assets	**19,800**	**18,984**	**19,037**	**20,019**	**21,296**	**17,235**
Diff (if any!)	0	0	0	0	0	0

Under international financial reporting standards (IFRS), profit & loss account and the balance sheet are designated "income statement" and "statement of financial position" respectively. UK-based companies are still using the old terms in their financial statements and the familiar terms are retained in this narrative.

2.5.2 *Optimisation of gearing*

As the gearing changes so does the NPV. The obvious reason why initially NPV increases with increased levels of debt is because of the greater relief on tax and because the cash flows are being discounted by the cost of debt, not equity. The lower cost of debt allows for a WACC lower than the cost of equity when the latter is determined from the standard capital asset pricing models (CAPM), which then significantly enhances the NPV.

The measure of the riskiness of a share is illustrated in Figure 2.6.

The beta (that is the gradient of the above plotted line) is a measure of a stock's volatility relative to the volatility of the market. A share that has a beta of 1.00 has the same return as the market.

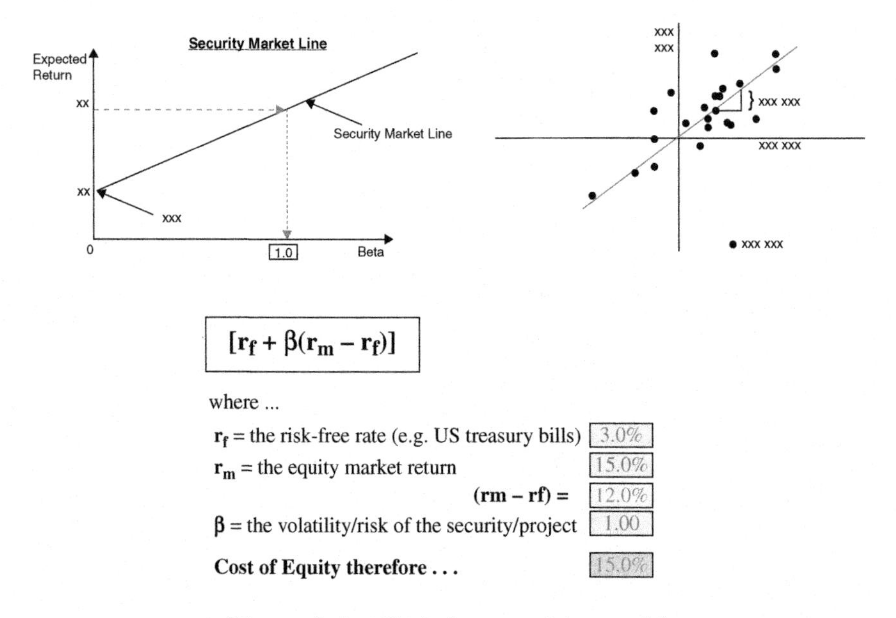

Figure 2.6. Capital asset pricing model.

Interest %	Tax %		DEBT	EQUITY	WEIGHTED AVERAGE
8%	30%				
1. After-Tax Cost of Debt			5.6%		
2. Estimated Cost of Equity				15.0%	
3. Debit/Equity Gearing			30.0%	70.0%	
4. Cost of Capital			1.7%	10.5%	12.18%

Figure 2.7. Determination of weighted average cost of capital.

If a beta is below 1.00 it is less volatile than the market and if above it is more volatile. The variables are then used to determine the cost of equity as indicated in Figure 2.6. Assuming the cost of equity is based on a gearing of 30% debt then at a tax rate of 30% and an interest rate of 8% the WACC is 12.18, as shown in Figure 2.7.

Any attempt to model gearing does obviously require that the cash flow model be integrated into the financial accounts, primarily so that the effect of tax can be correctly determined. Changing the gearing changes the risk and therefore the cost of equity and the cost of debt. The effect is shown in Figure 2.8.

The limiting factor is that as the level of debt increases so does the risk of default on meeting repayments on principal and interest. The CAPM recognises this and so the cost of equity increases. This will start off-setting the advantage of discounting by the WACC and NPV will, after reaching a maximum, start decreasing.

2.6 Petroleum Fiscal Regimes

A basic DCF model for a petroleum project will have many of the same elements as one set up for a mineral project. The project life will be derived from the production segments defined by annual production. The major difference is that, after a period of production in which volumetrics will be determined by the well infrastrucutre, there will normally be an exponential decline in production.

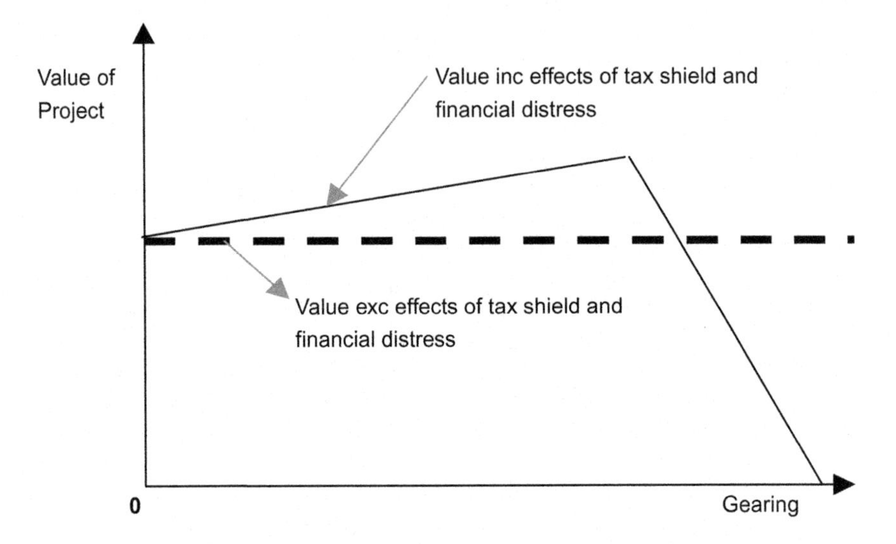

Figure 2.8. Gearing optimisation.

The volumetric used in setting up the model is based on a single value for the recoverable reserves. The production rate is then modelled over time using three segments — ramp-up, plateau and decline, each with their own input parameters. Bear in mind that there are an infinite number of combinations that can produce the same total field life volume. This needs to be linked to the decline rate in the final segment, so that the total volume produced matches the total reserves. The same result can be achieved by adjusting the plateau rate, plateau duration, ramp duration or the maximum of production periods in the model. The correct choice should be governed by reservoir performance, and it is not valid to simply squeeze or extend the production profile into the time available.

There is then the separate issue of the economic life of the project. This is termed the "economic limit", and is determined by the balance of revenues and costs over time. An example of a petroleum financial model is given in Figure 2.9 based on a linear decline curve, demonstrating the relationship between the economic limit and total operating costs.

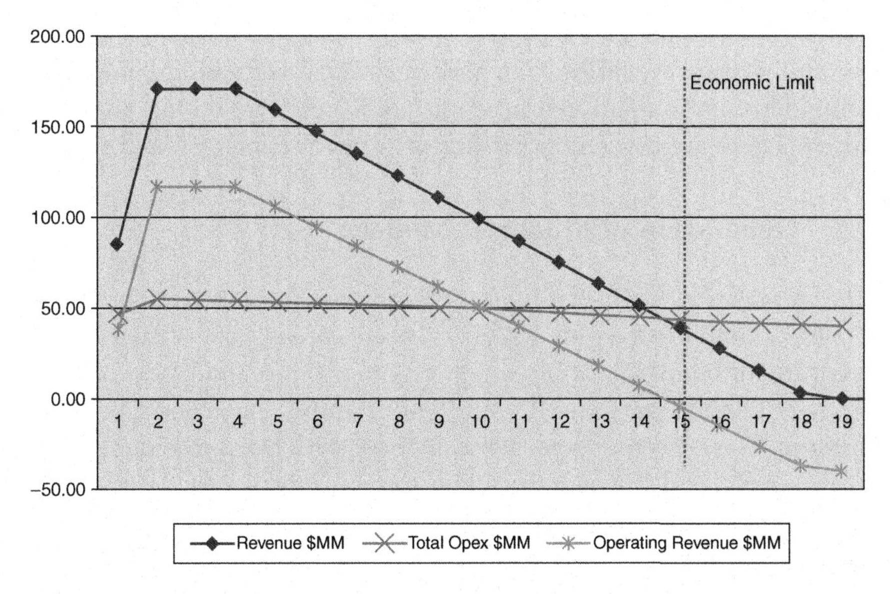

Figure 2.9. Cash flow for a petroleum project showing unconstrained and economic limit before tax cash flows.

There are three main types of hydrocarbon fiscal regimes. These are:

- Tax and royalty (concessionary) regimes.
- Production sharing contracts (PSC) or agreements, and
- Risk service contracts.

Of these the PSCs are the most complex. They are normally bespoke agreements specific to a field of jurisdiction and unintended consequences can arise once development and production commence (see Section 10.8). They are essentially complex tax structures which means that the relatively straightforward link falls away between tax-adjusted cost of debt used in mineral projects funded using debt. Choice of discount rate therefore tends to revert to the overall corporate cost of capital by the international petroleum company undertaking the development.

PSCs are modelled using revenue, cost recovery and profit split, and partner carry. The end result is an after-tax cash flow, which forms the fundamental basis for further analysis involving financing and gearing scenarios. Concessionary fiscal regimes are modelled using depreciation and hydrocarbon taxes, corporation tax and royalties. These are covered in further detail in Sections 4.5 and 10.8.

2.7 Generation of Financial Models

The application of IC-MinEval software was developed at Imperial College, London. It is an Excel^{TM}-based spreadsheet Visual Basic program automating all stages required to produce models for a wide range of mineral projects. IC-MinEval produces a balance sheet and profit and loss account from the cash flows, with tax provisions linked to the profit and loss account. The cost of debt is calculated, as is the weighted average cost of capital and the cost of equity. Output modules include the base case DCFs, as well as key financial ratios and performance indicators such as NPV, IRR payback and maximum cash exposure. Sensitivity analysis can be undertaken on key variables.

In 2000 the intellectual property incorporated in the development of IC-MinEval was transferred to an Imperial College start-up company called IC-FinEval Ltd and a coal version of IC-MinEval — IC-CoalEval — was developed and essentially incorporated the same functionality as IC-MinEval. This was followed by IC-Petroval which modelled cash flow models of oil and gas, projects based on volumetrics, segment production and annual production, revenue, opcosts, Capex, unconstrained and economic before tax cash flows, depreciation and tax and royalty, after tax cash flow and tax royalty and production-sharing contracts (revenue allocation, carry and cash flows). IC-Petroval remained as a beta version in the development stage and the application was never made available for general use.

The functionality of both IC-MinEval and IC-CoalEval was delivered over the internet through the Software as a Service (SAAS) system with InfoMine Inc (http://software.infomine.com/).

Chapter 3

Strategic Management

3.1 Introduction

This chapter considers the investment and value creation opportunities that are offered across the whole spectrum of the mining life cycle. The interrelationship between the life cycle and the drivers is illustrated in Figure 3.1.

A key driver in the minerals business is that natural resources have finite limits and therefore value creation is derived from an expansion of resources or reserves. An understanding of the life cycle of a project from exploration through to production helps identify those areas where value can be, and is, created.

The life cycle of an asset can be characterised by a number of stages. Each stage increases the level of certainty over the size of the potential resource base, the quality of the ore, and the likely costs and methods of extraction and benefaction.

As producing and development assets are often valued using a discounted cash flow methodology, reducing risk naturally increases the value. Earlier stage assets can be valued on a probabilistic method (expected present value) considering a range of likely payoffs, the capital expenditure required to gain those payoffs and the risk that expenditure will be incurred without success. Again, progression from exploration through evaluation to pre-production, utilising pre-feasibility and feasibility studies, decreases risk and enhances value through elimination of uncertainty over scale or quality of ore bodies and/or costs and timing of exploitation.

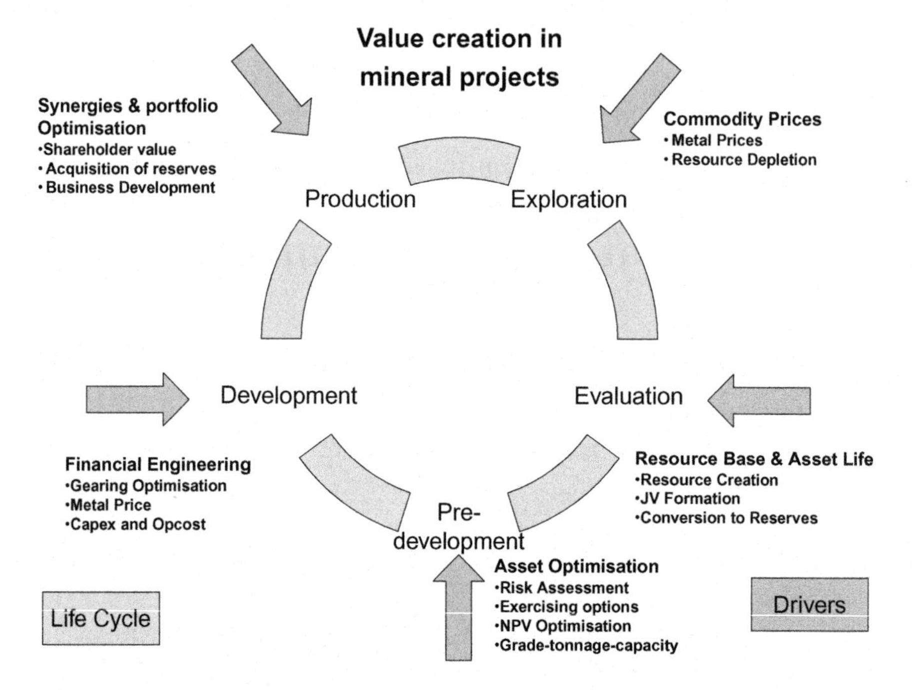

Figure 3.1. The cycle of value creation in mineral projects.

3.2 Exploration Stage

The proportion of targets identified at the exploration stage that will eventually emerge as operating mines is very small. The exploration and early evaluation stage is dominated by junior mining companies who generate value by screening areas and evaluating prospects. Given the relatively low success rate, the risk attached to such activities is high but junior mining companies make this exploration activity a core competence.

This is illustrated diagrammatically in Figure 3.2. The high rate at which targets are eliminated at the generative and exploration stages demonstrates the importance of maintaining a pipeline of targets.

Once the pre-feasibility stage is reached, the costs of commissioning an asset can exceed the funding available to the junior miner, which will need greater resources to exploit the asset. Larger mining groups are attracted to the fresh asset in order to enhance their

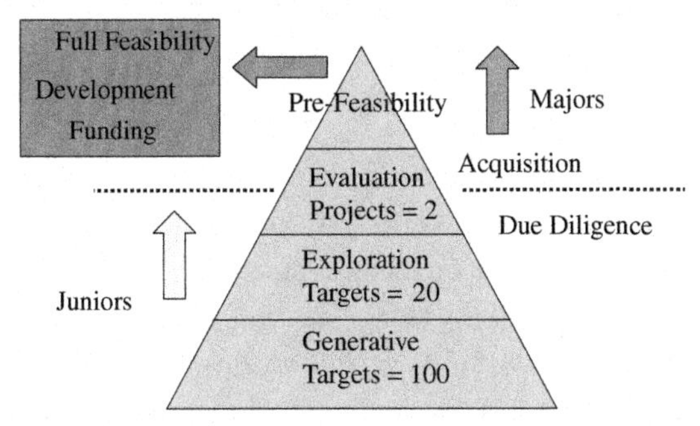

Figure 3.2. The pyramid reflecting the evolution of a gold mineral exploration and evaluation programme based on data from an exploration licence in Central Asia.

reserve base, and they are able to bring greater financial and technical capabilities to bear. Hence, the evolution of an exploration target into a mine will also often be accompanied by a series of mergers and acquisitions as the junior exploration company aligns with a larger mining group. This is a mutually beneficial arrangement allowing the mining groups essentially to undertake exploration by proxy, while from the exploration company's perspective entering into a joint venture with an established producer enhances their ability to raise equity funding. The need to create value by adding metal reserves to their underlying asset base is an important driver in encouraging major companies to invest in the exploration sector dominated by the junior companies. The higher risk associated with investment in early stage exploration is offset by the lower cost of securing a significant stake in the venture compared to the cost of a farm-in into more advanced projects.

3.3 Drivers

3.3.1 *Introduction*

Under either the expected present value or the DCF approach, there are a number of drivers that can be seen to impact strongly on value, regardless of volatility in commodity prices. As well as the obvious financial, geological and operating drivers, other aspects, such as entry to a new market, diversification of currency or commodity risk may also add value.

Use of broader metrics such as enterprise value (EV) — the economic measure that reflects the market value of a whole business as a ratio of earnings before interest, taxes, depreciation and amortisation (EBITDA) — or EV/production or EV/reserves need careful consideration as two projects with similar current earnings, production rates and/or reserves may not be equal. Differences may arise either in potential scale of the project (a pure production metric may ignore future growth plans), sunk costs (later stage projects may already have invested significant capital in removing overburden or on procurement of assets) or mine life (one asset may be closer to decline or require higher future costs to maintain production). These broader metrics are however much more useful in the valuation of corporate entities which may contain many projects in a range of stages of the life cycle.

3.3.2 *Commodity prices*

Whilst end commodity prices are an obvious driver of value they are outside the control of the mining entity, except where hedging is undertaken. Volatility in product price curves over the potentially long life of a mine, however, makes the revenue projections difficult. Operationally, this volatility impacts on the economics of the mine, decisions on optimising the design of the pit (as revenue estimates may change views on economic cut-off grades) and even the scale of development. For example, it may be worth scaling up quickly to gain access to higher short-term prices.

Additionally, volatility and recent high commodity prices have resulted in many major mining companies having low gearing and cash-rich balance sheets. These companies need to invest cash surpluses in extending reserves in existing operations, locate extensions to known mineralisation or acquire existing operations and known undeveloped deposits. Capital expenditure in the global mining industry has fluctuated over the last decade and has to be balanced against the need to generate profits. Boards know that they must either use their cash intelligently by developing new mines or they must return it to shareholders.

High metals prices of course also impact at an operational level. Locking in pay levels while commodity prices are high will, however, result in higher operating costs, leaving profits more vulnerable to any future falls in the price of the metals produced.

Companies and analysts often plan based on forecasts showing commodity prices falling (or rising) to historic averages. Even if historic averages represent the appropriate longer-term base level of prices, future spikes may increase returns in due course. Longer-term prices when demand and supply reach equilibrium are driven by costs of production. It is likely that the downside of below average prices is far exceeded by the potential scale of any spikes. There is a risk of conservatism in asset assessments and the inherent optionality in longer-term assets being forgotten.

3.4 Resource Base and Asset Life

3.4.1 *Mineral resources*

Successful exploration leads to evaluation involving complex geostatistical concepts of sampling practice and geological continuity, principles of uncertainty, spatial control on sampling, ore body modelling, grade-tonnage relationships and selection of cut-off grades. These are covered in Chapter 7.

Following the 1997/1998 Serious Fraud Office (SFO) prosecution of individuals involved in the Butte Mining flotation and the 1997 Toronto Stock Exchange (also known as the TSX) — listed Bre-X Minerals fraud, described by Jennifer Wells in her book of this title (see also Section 7.3), the professional bodies in the United Kingdom, Australia, South Africa and Canada undertook a major revision of the codes for reporting mineral resources. These are identified by their acronyms (for example, JORC[1] — Australian Institute of

[1] Joint Ore Reserves Committee.

Mining and Metallurgy — and SAMREC[2] — South African Institute of Mining and Metallurgy) and in plain English such as the Standards and Guidelines for Resources and Reserves — Canadian Institute of Mining, Metallurgy and Petroleum (CIM).

These professional organisations, together with the Geological Society of South Africa and the London-based Institute of Materials, Minerals and Mining (IoM3) jointly revised all their reporting codes in early 2000 and continually revise them. The IoM3's efforts evolved into the International Mining and Minerals Association (IMMa) and the Pan European Reserves and Resources Reporting Committee (PERC) Code. The mineral classification system is summarised in Figure 3.3. There is also the Committee for Mineral Reserves International Reporting Standards (CRIRSCO). This was formed in 1994 under the auspices of the Council of Mining and Metallurgical Institutes (CMMI), a grouping of representatives of organisations that are

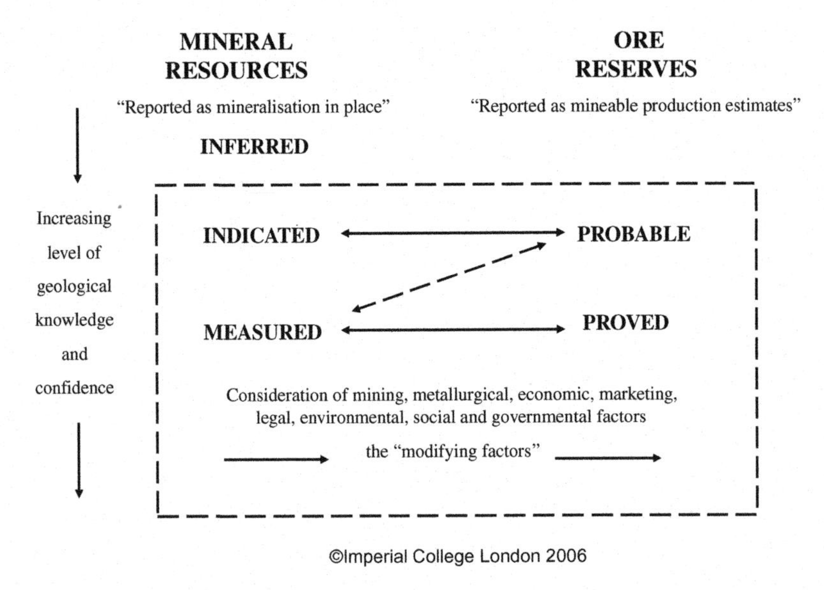

Figure 3.3. General relationship between exploration results, mineral resources and ore reserves.

[2]SA Code for the Reporting of Exploration Results, Mineral Resources and Mineral Reserves.

responsible for developing mineral reporting codes and guidelines in Australasia (JORC), Canada (CIM), Chile (National Committee), Europe (National Committee PERC), The Mongolian Professional Institute of Geosciences and Mining (MPIGM), Russian National Association for Subsoil Use Auditing (NAEN), South Africa (SAMREC) and the USA (SME).

The codes are essentially qualitative descriptions of uncertainty that are based on quantitative techniques. For minerals a resource category will be allocated by a geostatistician based on complex mathematical algorithms that combine concepts of statistical uncertainty with the spatial relationships between samples. Probabilistic concepts are simplified into deterministic estimates but it is up to the geostatistician to allocate the resource in mineral projects to the inferred, indicated or measured resource categories. Financial modelling also provides an objective basis for distinguishing between resources and reserves. Clearly as a project moves through the life cycle the ore body will move from resource to reserve and through inferred, indicated to measured. This progression is matched by value creation, as more certainty is gained over the quality and scale of the ore body and projects become financeable.

While these Codes were prepared for the equity markets and development projects they are now also being routinely applied to mature active mining operations. In these circumstances revisions to the definitions are normally introduced by the mine planning group as the level of sampling undertaken in an operating mine will always exceed that available for a development project (see Chapter 7).

There is a risk, however, that investors will assume that the level of sampling undertaken on pre-production projects needs to be comparable with those in operating mines exploiting similar styles of mineralisation, rather than that defined in the international reserve and resource codes.

Where inferred and indicated resources are very large, giving rise to potentially long mine lives, an NPV derived from a DCF model where cash flows are discounted at standard rates would not provide adequate weighting to revenues generated beyond 10 to 15 years. Indeed, given the impact of discounting, doubling the mine life may only add 10%–20% to the value. This tends to undervalue large projects even where a large probable reserve is demonstrated. Alternative strategies for assessing the valuations of these types of projects need to be considered. The application of real option analysis, that takes into account such variables as product price volatility, can result in valuations with a premium over the determined NPV.

In a climate of shortage of supplies there is every incentive to increase production capacity of long-life projects. Notwithstanding the need for additional expenditure on evaluation drilling and capital investment in the operation, the financial models may indicate enhanced NPVs for the same original resource based on the incremental production. Clearly it is not feasible to accelerate production in all cases, but scaling operations may add significant value.

Recent opinion is that more importance should be given to inferred resources even if they are not being given more weighting in valuations. Promoters are now being encouraged to generate "what if" DCF scenarios even though the Codes restrict the application of economic modelling to indicated resources. This is based on the premise that the potential for a positive NPV must to be demonstrated if the cost of closer spaced drilling needed to take the resource from the inferred into the indicated category is to be justified (see Chapter 7).

3.4.2 *Petroleum resources*

The corresponding reserve classification for petroleum reserves is given in Figure 3.4.

Probabilistic reservoir estimation techniques and simulation technology are used to determine the underlying key relationship between "total petroleum-initially-in-place" and "estimated ultimate recovery". Most oil and gas analysts understand that the 1P reserves (proven only), which the US Securities and Exchange Commission (SEC) require, are no basis for valuing the future earnings potential of a company. They will look at risk-weighted 2P (proved + probable) reserves (see Section 4.6). Companies are mostly interested in ensuring they have a rigorous, auditable system to support their estimate

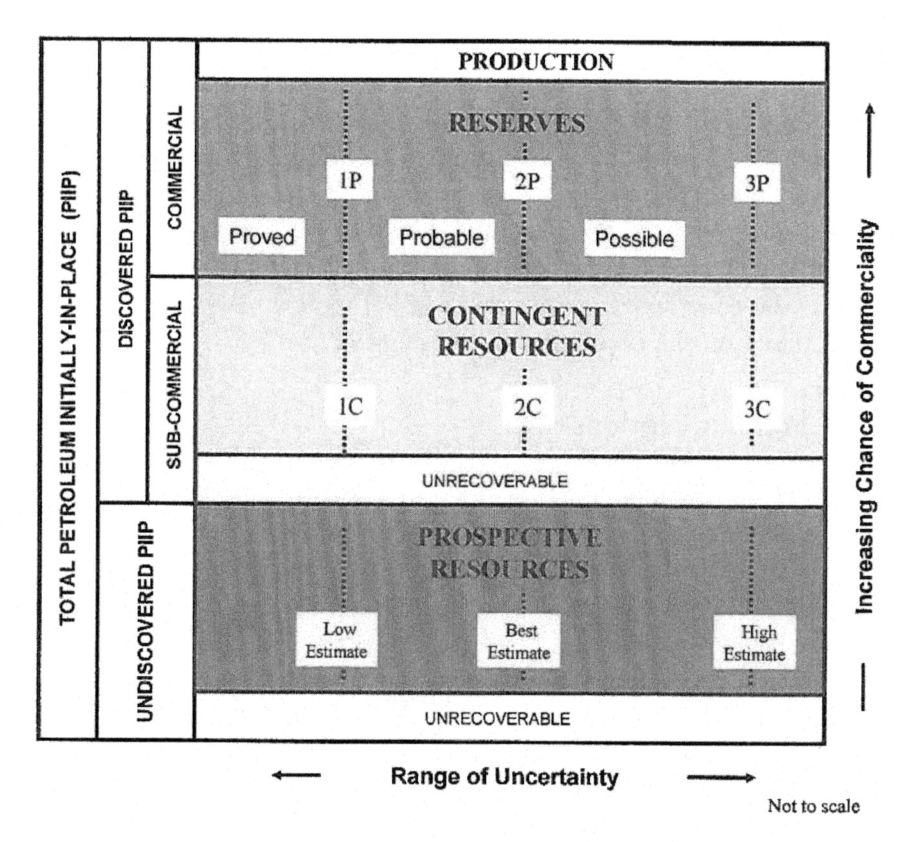

Figure 3.4. Society of Petroleum Engineers reserve classification: petroleum resources management system.

of the reserves submitted to the SEC. Directors won't get sued for taking a calculated geological risk, but they will get sued for not having a system to back up their numbers.

While there are obvious similarities in the resource and reserve nomenclature in petroleum and minerals, direct comparisons are probably not possible. The approach to extraction is of course also fundamentally different.

Energy projects such as coal, uranium and oil sands are essentially mining projects and can be accommodated within the Minerals Reserve Code.

The NPVs of all oil and gas projects were historically based on a simple DCF using a standard discount rate of 10% (designated PV10) for SEC-reporting purposes. Companies are probably happy to report normalised NPV10 numbers to SEC. This convention is, however, rarely the basis of share price valuations. Predictions of the price of petroleum company shares should be derived using the same gearing considerations as used in the minerals industry. In reality, due to the complexity of petroleum fiscal models, DCF models are often based on the corporate cost of capital.

3.5 Funding Options

A listing on the main LSE is not possible for companies involved only in exploration — the company must have at least one project associated with a probable reserve. It is an option that is restricted to companies undertaking or proposing to undertake extraction on a commercial scale.

For projects evolving into the development phase, project finance becomes an option. Banks exposed to the level of risk associated with a mining project will naturally take a conservative view. The development phase of a project would normally involve a joint venture (JV) with experienced operators.

Once projects have advanced to the stage of a full technical feasibility study, valuations determined from financial models based on a simple discount rate and a probable reserve will tend to approach the full potential NPV of the project. For a single project company this would also tend to be the same as its market capitalisation.

Clearly volatile commodity prices increase tensions between miner and financier. Nevertheless, in the absence of a JV with a larger miner (with access to traditional corporate finance based on lending capacity across its portfolio), a junior miner will have to risk losing upside in order to secure the finance needed to progress the project.

3.6 Synergies and Portfolio Optimisation

To expand beyond its current reserve base the major mining group will normally either have to buy an existing deposit or discover a new one. As many major mining houses have delegated exploration to the juniors, new reserves will generally be built via acquisition. If the type of deposit sought has to be large and high grade then it is probably blind (there is no surface exposure) and will be very difficult to delineate or very expensive to acquire if already developed. Nevertheless there is a major drive by the majors to acquire companies with existing reserves.

Past major acquisitions have shown that value can be created by mega-mergers even at the top of the commodity cycle. Some corporate acquisitions involved multiples of NPV being paid. But how can value be created if acquisition cost exceeds that determined using a conventional DCF-based valuation? Portfolios of projects may, however, be value creative beyond plain vanilla DCF, from a corporate perspective, through elimination or diversification of currency, country or commodity risks.

Additionally in the corporate context, significant cost savings may be achieved, longer-term credit ratings improved (giving access to cheaper bank financing) and new competencies or enhanced reserve bases acquired. In this respect, valuing a miner based on the sum of the parts of the NPVs of its stand-alone projects may be erroneous. The NPV of the combined corporate holdings could easily exceed that of the underlying assets owing to the inclusion of future growth in blue-sky projects. The corporate holding in this context provides investors with a vehicle to invest in both the underlying commodities but also a management team with proven abilities in growing the business. In this context the use of EV/EBITDA multiples becomes more useful. It is not just the current earnings and reserves that are important, a premium may be warranted for either organic or acquisitive growth. It is this momentum and the desire to maintain it that may partially explain the seemingly high prices that were paid for mineral projects prior to 2008. Another explanation for a multiple of NPVs arising is recognition of the conservatism that often pervades mineral and project assessments. If original technical, geological and commodity price assumptions are all set at conservative levels and then a low or normal discount rate applied, prices paid for assets will appear as a multiple of NPV as more optimistic scenarios are modelled.

3.7 Value Creation in Mineral Projects

Recent volatility in metal prices and its impact on the mining sector has increased the need to understand the interrelationship between technical and financial risk. The key consideration is the intrinsic value of a mineral project. Where current market conditions undervalue the true long-term worth of assets, which is then reflected in low share prices, strategic planning can identify opportunities. The following sections review strategic approaches for evaluating projects at the prefeasibility stage and identify the underlying financial and technical principles which apply to mineral projects. Particular attention will be paid to the treatment of the key independent variables, such as grade, and dependent variables, such as grade-tonnage relationships and the way these influence the rate of mining, associated costs and optimisation of the net present value of a project.

Professionals in the investment community appreciate that the higher the risk the greater the opportunity for significant returns on an investment in the sector and the leveraged position gained in the underlying commodity gained via an investment in a natural resource business, whether minerals or petroleum. This is illustrated in Figure 3.5.

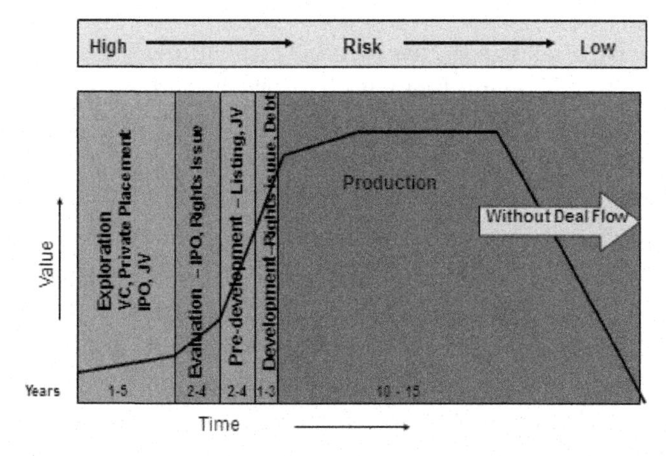

Figure 3.5. Single project mining company stages of development and funding options.

Key boundaries on the value added line are the inflection points when exploration transitions into evaluation and evaluation into pre-development. The experienced investor might reason that once production starts that is the time to sell down a holding as the reality of mining becomes evident. Figure 3.5 is of course simply a generic view in terms of funding. A major mining company could fund all the stages from internal resources.

3.8 Funding Options for Mineral Projects

3.8.1 *Pre-initial public offering (IPO)*

For the junior mining company with largely exploration assets, raising equity on a stock exchange (also referred to as a "listing") funding options are often limited to private placements, with valuations very difficult to establish and often highly subjective. This does, however, allow the investor to link the higher risk associated with investment in early-stage exploration with the lower cost of securing a significant stake in the venture compared to more advanced projects.

For a pure exploration project the best option is to establish an unlisted company with backing from a limited number of private investors who have a major stake (shareholding). These are either venture capital funds or private placements. There is no liquidity for the shares they hold. This is the first step for any entrepreneur. See also Section 8.2.

3.8.2 *Listing*

The prospectus for a listing must have a technical report, prepared by an independent consultant (Competent Person's Report), which in turn may include all the elements of a pre-feasibility study. It may not be necessary to have a full technical feasibility study completed (see Chapter 8) if funds are to be used for, say, additional drilling or trial mining.

A listing on one of the main exchanges such as the LSE is not possible for companies involved only in exploration — the company must have at least one project associated with a probable reserve. Once the target generation/exploration stage has been completed under a reconnaissance permit and the company has confirmed the presence of significant mineralisation, a public listing might be possible on one of the specialist exchanges such as the LSE's Alternative Investment Market (AIM) or the TSX's Venture Exchange (VE). You cannot generate a DCF or undertake a preliminary feasibility study (PFS, see Section 8.3) without an indicated resource according to JORC, SAMREC, CIM, etc. codes. Once you have achieved that you could expect to generate an NPV of \$200–\$500 million.

Once you have decided that this is the opportune time to go public and list on a public market, you prepare a prospectus based on AIM, The Australian Stock Exchange (ASX), TSX, etc guidelines which includes the Competent Persons/Qualified Persons Report. Clearly this is a pointless exercise if it is not linked to attracting investors. Investors are not going to come to you, you have to approach them with a compelling reason why they should invest in your company.

If you conclude that you need \$30 million for the next stage of evaluation drilling then that is what you must raise. You create say, 160,000,000 shares at 25 cents per share: 120,000,000 will be issued.

No fund manager is going to take up all the shares on offer but for reputable specialist commodity funds (SCF) it is hardly worthwhile investing less than \$6 million which would give them 20% of the company. They may object and say that for \$6 million they want 30% of the company. You will then need either to lower the price of the share or issue more shares to the fund. You then move on to the next fund manager saying that a well-known fund is taking 20% and of course you are pushing against an open door. The word may then get out onto the street and private investors have the opportunity to apply, but this would be only a small proportion of funds flowing to a new natural resource IPO.

You then list and on the first day of trading your share price goes up to 30 cents. Everyone is happy and your market capitalisation is \$36 million. You got your pricing about right. That is the value of your company. You then start spending the money on your drilling, achieve the milestone outlined, generate an NPV of, say, \$200 million following a PFS based in part on metallurgical tests and preliminary engineering — the market believes this and so your Market Capex reflects this number. Your SCF sells out, say, half of its 20% stake for $20 million having now made $14 million for its shareholders.

3.8.3 *JV Agreement*

Having identified a deposit potentially able to support a commercially viable operation the Directors will then need to consider how you are going to get into production. This is when reality sets in — pyro- and hydrometallurgical technical issues around downstream production of, for example, platinum group metals (PGE) from an initial concentrate are highly complex.

The development phase of a project would normally involve a JV with experienced operators. Putting together a JV requires a valuation of the assets of the junior as a basis for the determination of the vend-in conditions for the major, and a clear set of targets to identify when trigger points are reached. Usually this is linked to the delivery by the major of a pre-feasibility and a full technical feasibility study. These need to be defined with rigour if disputes are to be avoided. In the event that the major decides not to proceed, a claw-back clause is often included in the terms of the original JV agreement. Announcing the JV will of course be good for your share price.

The vend-in conditions will be dependent on the valuation you have managed to achieve and this has to be linked to market capitalisation if the company is listed. The gold standard for a valuation be able to generate a DCF analysis. This requires, however, that the project be associated with an indicated resource. Even for advanced-staged exploration this milestone may not yet be present. Comparable transactions require identification of market capitalisation of several junior companies with licences covering similar metallogenic characteristics. The appraised value method assumes that a property's value is either enhanced or diminished by an exploration programme and that funds expended will produce value which will be proportionate to the expenditure. Of course if drilling sterilises a potential target then the accrued costs need to be written off, so additional expenditure on a prospect can end up reducing the valuation. Intangible assets can be added to valuation determined

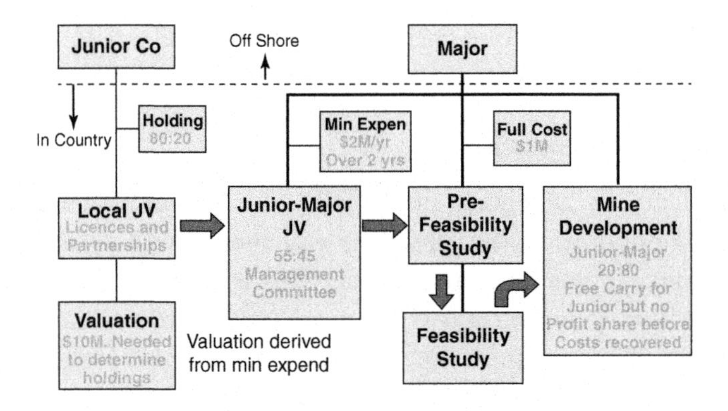

Figure 3.6. Components of a joint venture.

using the above methods. These include local infrastructure, deferred expenditure and compliance with minimum expenditure requirements.

The basic structure of a typical JV agreement is given in Figure 3.6.

As major miners often explore by proxy via JVs with junior miners, the alternative strategic approach to acquiring mature developed assets is to vend into a large project post-discovery but pre-development. At this point there is still some project uncertainty but the major has the ability to utilise experienced project development staff to accelerate the project and reduce project risk. This gives the ability to create value within the project itself as well as through the addition of reserves to complement the major's existing operations.

A major international mining company seeking to enter into a JV with a junior company holding a prospecting right where a base metal exploration project has entered the evaluation drilling stage needs to take into account the following:

- The profile of the junior.
- The merits of the property and project.
- The structure of the likely business relationship.
- The legal framework of JV agreement, and

- The relationship between the vend-in expenditure, the valuation of the project and the target shareholding to be acquired.

The earlier the major can vend in, the less they pay for a significant shareholding. At the same time the more the junior concedes as a shareholding the greater the dilution for the junior's founders. This effect may be offset by the enhanced share price arising from a successful JV. The assumption is that the major will use their funds to supplement the junior's cash to convert a larg(ish) inferred resource into an indicated resource to allow a pre-feasibility and then possibly a full technical feasibility to be prepared.

Figure 3.6 should also be considered in relation to Figure 10.6 showing a tax-efficient company structure. In a perfect world the junior company would have a corporate structure such that expensive re-structuring is not needed if an exploration programme is successful and a significant discovery made. In the real world legacy agreements with local strategic partners and sources of funding mean that restructuring is often needed.

3.8.4 *Project finance*

For projects evolving into the development phase, project finance becomes an option. Banks exposed to the level of risk associated with a mining project will take a conservative view. They generate an information memorandum which incorporates a full technical feasibility study. The factors involved include the following:

- Metal price assumptions and hedging.
- Capital, opcost, contingency and reserve tail assumptions.
- Sensitivity analysis and role of tax.
- Gearing performance indicators.
- Gearing optimisation, and
- Multi-partner modelling.

This is discussed in more detail in Section 8.6.

3.8.5 *Discussion*

Estimates of the grade and tonnage in the full technical feasibility study may be no different from those used for pre-feasibility, but may well include the results of trial mining, bulk sampling and pilot treatment. This level of technical information would not usually fall within the scope of a pre-feasibility study. Permitting and environmental impact assessment (EIA) (see Section 4.1.1), also need to be in place. A full technical feasibility study will also provide the verifiable cost and performance predictions. These are needed for the economic completion tests when operations funded from debt progress from the recourse to non-recourse stage. These issues are discussed in Chapter 8.

Major international mining companies act virtually as their own investment banks in setting cost of capital. Projects evolving into the development phase therefore provide a favourable investment environment for a rights issue, exchange-listed convertible bonds, high-yield notes and structured corporate loans.

High commodity prices also provide a platform from which debt can be quickly repaid. Strong cash flow enables acquisitions to be debt-funded and gearing reduced as the acquisition is digested. Reasonably rapidly, the acquirer is able to un-gear its balance sheet or that of the target, leaving capacity for further acquisitions or greenfield investment. When there is an appetite for mining stocks this gives the major the option of raising new equity, or even funding from cash flow.

Chapter 4

Management of Projects, Markets and Supplies

4.1 Constraints on Mineral Resource Development

4.1.1 *Environmental impact assessment*

There is a disconnect in society between the necessary requirement to minimise the environmental impact of mining during production, including the obligation under a mining licence to fully rehabilitate the site once mining has ceased, and the nostalgia for historical mining districts. An excellent example of the transition from industrial dereliction to historic heritage are the 19th century pump houses of Cornwall built during the peak of the tin and copper mining period which form the backdrop to the BBC *Poldark* television series. Mining also has an archaeological context. The Las Medulas gold mine in Spain was developed by the Romans with description given by Pliny the Elder, Gaius Plinius Secundus (AD 23–25 August, AD 79) in Book XXXIII of his *Naturalis Historiae.* The massive excavation was created through the extensive use of hydraulics with viaducts extending up to 100 km from the site. The alluvial gravels were then washed into gorse-lined sluices and gold recovered after the gorse was dried and burnt. This is now a World Heritage site.

Prior to mining, an EIA must normally be undertaken. This would include consideration of surface water, groundwater, air quality, soils, land use, socio-economic issues, human amenity (noise, vibration and traffic), archaeology, landscape and visual impact. Each of these aspects would in turn need baseline conditions and potential impacts to be determined, and a strategy to mitigate impact put forward. This would involve provision for environmental management and monitoring. The scale of the task and the need for an interdisciplinary approach means that the work is best undertaken by a specialist environmental services company who employ biologists, engineers, social scientists, archaeologists, hydrologists and many other specialists. They in turn will need to coordinate their work with consultants undertaking the detailed engineering design stages (see Section 8.5.3) who may need to accommodate constraints required for the mining licence to be awarded. The siting and design of tailings dams is often a particularly contentious issue (see Section 9.5.1).

4.1.2 *Sustainable development and the moral case for mining*

Given the global reach of natural resource development there are international guidelines on their development.

The Equator Principles are a risk management framework adopted by financial institutions for determining, assessing and managing environmental and social risk in projects, and are primarily intended to provide a minimum standard for due diligence to support responsible risk decision-making.

The World Bank provides guidelines for the development of base metal iron ore and coal mining as well as copper smelting and steel manufacturing.

The International Council on Mining and Metals (ICMM) was established in 2001 to act as a catalyst for performance improvement in the mining and metals industry.

The Extractive Industries Transparency Initiative (EITI) is a global standard to promote open and accountable management of natural resources. Stakeholders include all the international mining and oil companies. It seeks to strengthen government and company systems, inform public debate and enhance trust. In each implementing country the initiative is supported by a coalition of governments, companies and civil society working together.

The question that has to be asked is to what extent does an international mining company, when operating in a developing country, have an obligation to ensure that the revenues accruing to the host government are spent in a way which brings about broad, sustainable economic development and are not wasted through mismanagement and corruption? What are the obligations of a mining company to influence the behaviour of the government in a host country to ensure an equitable distribution of benefit flowing from the development of large projects? Should they demand economic and political reforms as a condition of their investment, or adopt a "non-interference" strategy (as Chinese mining companies have tended to do)?

What are the risks of undertaking mining extraction of natural resource projects which represent a significant proportion of a host country's gross domestic product (GDP)? Is there a case for applying the approach used by the petroleum industry based on a PSC rather than a tax and royalty regime which is likely to reduce political sensitivities (see Section 2.4)? Governments know perfectly well that with a well-designed PSC, the income to the treasury could be exactly the same as a tax and royalty regime.

Where major infrastructure projects are needed to release value from bulk commodity deposits such as iron ore or coal projects, the role of the International Finance Corporation (IFC), part of the World Bank Group, becomes important as a moderating influence. The European Bank for Reconstruction and Development can have a role in addressing corporate governance issues.

4.1.3 *Occupational health and safety*

Given the key role that good management has in a mining operation, any objective assessment will need a site visit. The first expectation is that there should be evidence of good housekeeping. If the site is a mess, then that is visible evidence that it is poorly managed. An untidy workplace is also inherently unsafe. Operators will have to deliver what the manager demonstrates through action is really required, not what is said is needed. This impacts on safety: a manager can talk about safety but will be ignored if, on inspection, he passes unsafe situations without comment. On a visit one should be able to tell whether all staff respect the manager. If they do then the person is clearly doing things right. The operation will only really give priority to safety when the manager refuses to pass an unsafe situation until it is fixed. There are no unsafe mines, only ones with unsafe management.

The same is true about other aspects of a mining operation. This is one of the reasons why so many waste tonnes end up in the plant — most managers really want to get their daily tonnage come what may. It is very rare to have a manager who worries more about how many ounces of gold arrived at the plant rather than how many tonnes of ore.

In the final analysis, if the management really wants a clean, tidy, safe operation they will get it. It will also almost certainly be a highly efficient one as a result. Needless to say the situation with a poorly managed operation will not be improved by abstract analysis, and investing further funding could simply reward complacency.

4.2 Cost of Environmental Compliance and Closure Provision

The cost of meeting the requirements for compliance with environmental constraints might start with comparing the NPV of two identical projects — one with permitting in place and the other, say, two years away from this stage. A DCF model of the former would generate a significantly higher NPV as the revenue stream would be brought forward. While it might imply that the difference between the NPVs of the two projects is the cost of complying with environmental constraints, this would be wrong. It is simply an artefact of DCF modelling. For this reason it is recommended that financial models should assume that permitting is in place.

Environmental compliance can also generate an artefact in a DCF model when there is a significant cost of rehabilitation at the end of the mine life where, counter-intuitively, increasing the discount rate can improve NPV.

A DCF model generates results that reflect the true financial impact of compliance with permitting requirements for reducing environmental impact during production and undertaking site rehabilitation after closure once a mining permit has been awarded.

These compliance costs include the following:

- Bullet payment at the end of the mine life. This is a fixed amount which is payable at the end of the mine life for the environmental rehabilitation of the mine site, retrenchment costs, etc. As this is discounted in a financial model it is usually not material but there have been examples of projects which have such huge clean-up liabilities that the NPV improves as the discount rate increases. This counter-intuitive outcome arises because the liability is being reduced in the model.
- Environmental sink fund at beginning of production. This is usually an alternative to the above and is a fixed amount which is payable before the start of production and acts as an environmental bond to cover the cost of the environmental rehabilitation of the mine site. This applies to a country such as Ireland, which has a legacy of environmental dereliction which the state has had to deal with. This cost is undiscounted, so will reduce NPV by the equivalent amount.
- Annual environmental costs during production. A well-designed mineral operation should have provision for rehabilitation built in as an integral part of the cycle of extraction and disposal of waste rock and tailings. For example, on the mining side double handling should be avoided. On the processing side waste rock should not be mixed with sulphide-bearing low-grade ore and chemically unstable sulphide in tailings should be kept in a reducing environment (see Section 9.6.1).Under these circumstances environmental compliance should not add significantly to base case operating costs.
- Annual rehabilitation costs after mining. This is the annual cost of rehabilitating the mine site and is not included in either the sink or bullet payments that have been made. A modern project is unlikely to gain a mining licence unless there is a clear plan for closure and site rehabilitation. It will therefore have been designed with this in mind so the cost should not be disproportionate to the cost of development.
- Number of years for environmental clean-up. Essentially the length of time after the completion of mining that the annual rehabilitation costs have to be paid. For the reasons given above

this should not exceed two years. This is different from the requirement in many countries that there should be site monitoring after closure and rehabilitation. In the European Union this period is essentially in perpetuity. The bullet payment at the end of the mine life may incorporate some type of annuity to fund the process of long-term site monitoring.

Financial modelling can, perversely, be used to demonstrate that the best way of preserving value for future generations is to leave natural resources in the ground where they will remain safe. Net future value always exceeds NPV. The trick is to find a mechanism, such as used by the Norwegian Petroleum Fund, which allows sustainable wealth from the current exploitation of a natural resource to be preserved through fiscal modelling.

4.3 Site Visits and Due Diligence

The formal due diligence provides the basis for significant investment decisions. The guidelines are as follows:

- Objective — Ensure that all material technical, financial and legal issues are disclosed (the responsibility of the promoter) and reviewed (the responsibility of the evaluator).
- Obligations of the promoter — Set up a data room. Provide administrative and professional support. Select a set of relevant reports. Provide a digital (access) database listing of all reports. Arrange site visit. It is counterproductive to overload evaluators with superfluous information. Promoters are not obliged to identify risk elements where these can be deduced from the disclosures.
- Obligations of the evaluator — Form an inter-disciplinary team. Assign a suitable amount of time for a site visit (normally around one week). Keep a record of material examined and copied. Formulate questions.
- Preparation by the evaluator — Be sure of the reason for the visit and collect all currently available information on the site. Study the available information and question any unclear items — ensure the current status is clear. Prepare a list of items that need to be seen

on the site visit, particularly items critical within the constraints of the purpose of the visit. Inform site personnel of items you wish to see that may otherwise be missed on the visit, and
- At site — Make the purpose of the visit clear from the start. Look for tidiness — a tidy site is almost always an efficient site. Check safety records — there is nothing as unproductive as an accident! Look at the control mechanisms — are they fit for purpose and properly maintained? Make sure to see all the items on the checklist prepared for the visit.

The evaluator on an assignment has to formulate the right questions. Get this wrong and you will simply demonstrate a basic ignorance in understanding of which the promoter will take full advantage. Of course it works the other way around too.

If you are promoting a project to an evaluator you want to be able to assess their competence. Post Bre-X (see Section 3.4.1) the integrity of the sampling information should be absolute. Check if in doubt. Gold and diamond projects are particularly vulnerable to salting. Retaining the integrity of technical results in a corrupted programme is, however, very difficult for anyone perpetrating a fraud — a process mineralogist after all reported alluvial gold in Bre-X samples which, for a purported epithermal deposit, was a clear red flag (see Section 7.3).

4.4 Petroleum Fiscal Models

The approach to financial modelling needed to undertake valuations of petroleum assets is given in Section 2.4 and developed further in this chapter. The approach uses the same principles as the valuation for minerals projects in that you start with basic technical assumptions (which of course will be different for a petroleum project). The fiscal factors for a petroleum project can be a lot more complex than for a mineral project. They comprise the following:

- Tax and royalty (concessionary) regimes. These are similar to those in mineral fiscal regimes. The title to the resources is owned by the licence holder and stakeholders bear costs, receive revenues,

		Company Share	Government Share
Gross Revenue (Oil Price)	$60.00		
Royalty	12.5%	$52.50	$7.50
Assumed cost	$5.65	$46.85	
Special petroleum tax	25%	$35.14	$11.71
Income tax	35%	$22.84	$12.30
Division of Cash Flow		$28.49	$31.51
Take		42%	58%
Lifting entitlement		87.5%	12.5%

Figure 4.1. Tax and royalty system flow diagram showing distribution of revenue from one barrel of oil.

			Company Share	Government Share
Gross Revenue (Oil Price)	$60.00			
Royalty	10.0%		$54.00	$6.00
Assumed cost	$5.65		$48.35	
Special petroleum tax	25%		$35.14	$11.71
Income tax	30%		$19.34	$29.01
Profit oil split contractor	40%		$13.54	$5.80
Profit oil split government	60%			
Division of Cash Flow			$19.19	$40.81
Take			25%	75%
Lifting entitlement			42%	58%

Figure 4.2. PSC flow diagram showing distribution of revenue from one barrel of oil.

and pay royalties and taxes in proportion to their equity. This is illustrated in Figure 4.1.

- Production sharing contracts and agreements. Title to reserves always held by state, or its proxy and the "contractor" and state enter into contract to explore for, and develop, reserves. The contractor bears all costs (and therefore risks), and is paid for services out of the project revenues. The contractor is paid via two key mechanisms: cost recovery and profit sharing, as illustrated in Figure 4.2, and

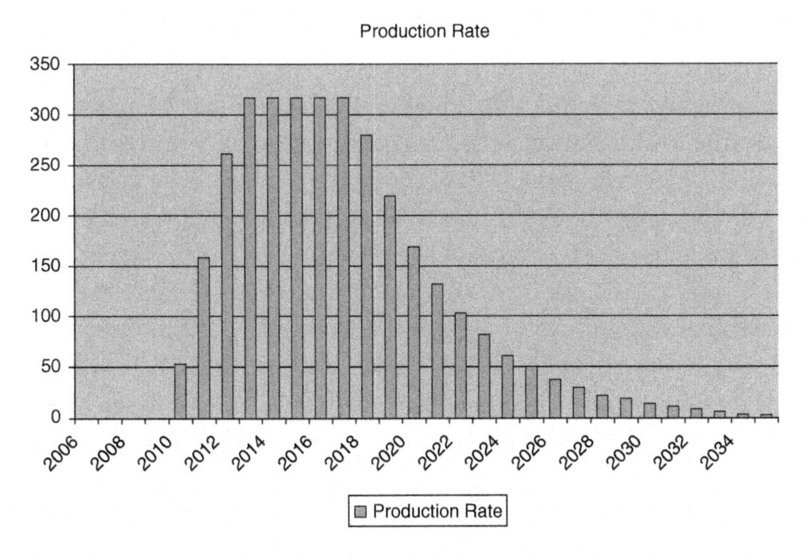

Figure 4.3. Petroleum projects are characterised by decline curves.

- Service contracts do not give an ownership right to oil in the ground and the company is simply paid a fee for its services in extracting the government's oil.

The typical production profile of a petroleum project is illustrated in Figure 4.3. As production declines, so revenue declines. The point in time at which the revenue has declined to equal the operating costs is known as the economic limit of the field, as illustrated in Figure 2.6 and discussed in Sections 2.4 and 10.8.

4.5 Petroleum Resource Management System

An oil reservoir requires a rock to be both permeable and porous. Permeability measures the ability of fluids to flow through rock (or other porous media). Typical values of permeability for petroleum reservoir rocks range from 10,000 to 100 millidarcies. Porosity is a measure of the volume of voids which cannot exceed 45% (see Section 6.2).

Natural production occurs when the reservoir pressure exceeds the well pressure. This is due to a combination of hydrostatic pressure at the base of a column of water exceeding that for an equivalent column of oil due to the density difference. This effect reinforces lithostatic pressure, which is much higher than hydrostatic pressure, and means, due to the density of sedimentary rocks (2.6 g/cm^3), that once breached by a well, oil will flow up to the surface with considerable force. Drilling technology must take this into account and blow-out protection is an integral part of the design. The Deep Water Horizon oil spill and the Macondo blowout are testament to the environmental implications of a failure of systems in the petroleum industry.

Once pressure in the reservoir equals well pressure, flow will cease, which may occur after as little as 10% of the stock tank oil-initially-in-place (STOIIP) has been recovered. Secondary production is maintained by injecting gas or water but typically as little as 35% of STOIIP may be recovered.

IC-Petroval incorporates a series of forms that provide basic input variables needed to calculate volumetrics, as illustrated in Figure 4.4.

IC-Petroval will model the production rate over time using three segments — ramp-up, plateau and decline, each with their own input parameters. Bear in mind that there are an infinite number of combinations that can produce the same total field-life volume.

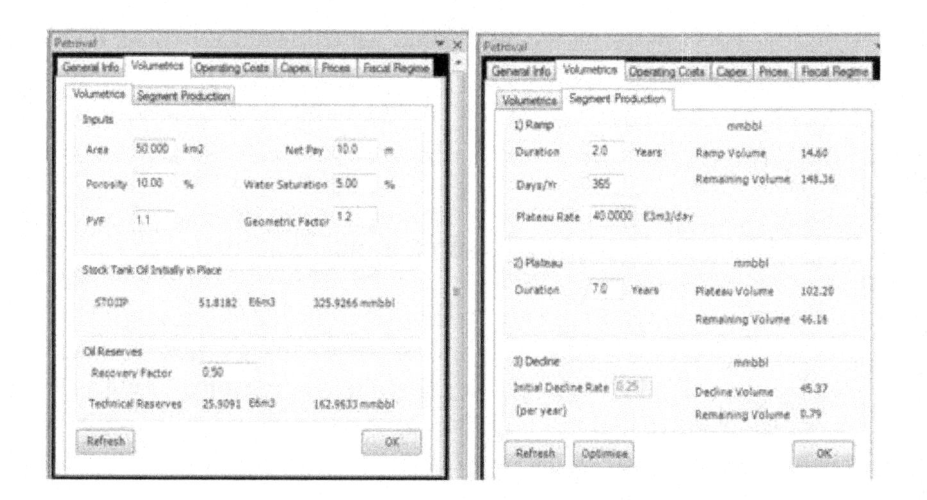

Figure 4.4. Volumetrics and segment production forms used in IC-Petroval.

The simplest approach for calculating recoverable reserves and their associated uncertainty is MC simulation. STOIIP is the amount of oil that is estimated to be present in the reservoir. STOIIP may be multiplied by an estimate of the recovery factor in order to estimate recoverable reserves. This is derived from the following:

- Gross rock volume (GRV).
- Net-to-gross ratio (N/G) which is the ratio between the volume of productive (net) reservoir rock and the total (gross) reservoir volume.
- Porosity ($\boldsymbol{\Phi}$).
- Oil saturation (So), and.
- Formation volume factor (FVF) which relates the volume of oil at stock-tank conditions to the volume of oil at elevated pressure and temperature in the reservoir. Values typically range from approximately 1.0 bbl/Stock Tank Barrel (STB) for crude oil systems containing little or no solution gas to nearly 3.0 bbl/STB for highly volatile oil.

$$\text{STOIIP} = \text{GRV} \times \text{N/G} \times \Phi \times \text{So} \times \text{FVF}.$$

A MC simulation can be generated from the assumed probability density function (pdf) of these five variables in order to generate the histogram given in Figure 6.2 (in this Crystal Ball simulation 100,000 samples of each parameter were generated probabilistically).

Crystal Ball simply calculates the P90 (resp. P10) as the value above which nine-tenths (resp. one-tenth) or 90,000 of the samples (resp. 90,000) are found. P90 and P10 in Figure 4.5 are $0.13\,\text{mm}^3$ and $0.51\,\text{mm}^3$ respectively. Since the histogram above is clearly lognormal, P90 and P10 could also be calculated from the theoretical quantiles of the lognormal pdf. The Central Limit Theorem of probability theory implies that the product of random variables tends to be log normally distributed.

Confusion often arises because the nomenclature of P10 and P90 are switched around as compared to the quantile nomenclature of statisticians. What is a q10 (10% quantile) for statisticians is a P90 for petroleum engineers. It means the value at which there is 90% probability that the volume is greater than this value. The same applies to q90 vs P10 on the other side of the curve.

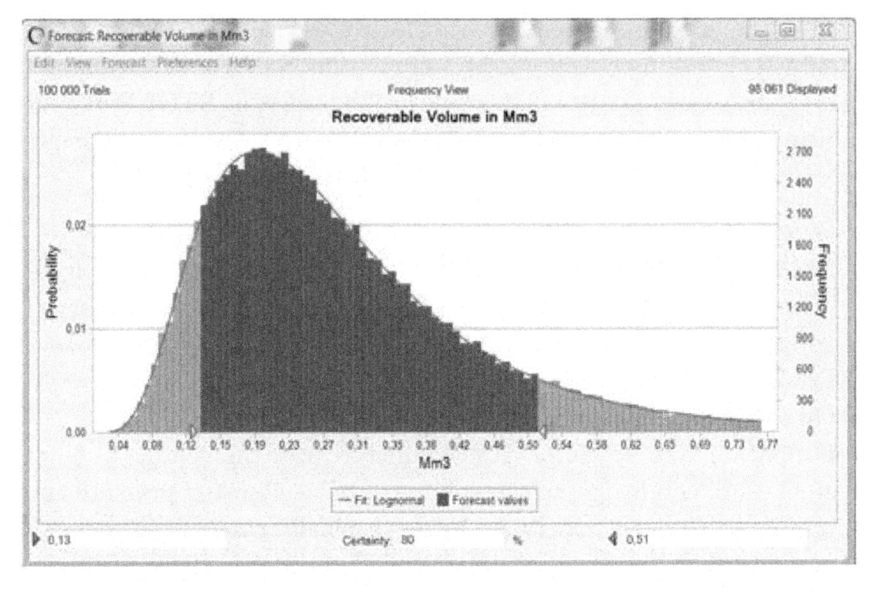

Figure 4.5. Positioning P90 and P10 on the output histogram generated from a MC simulation using Crystal Ball©from gross rock volume, net-to-gross ratio, porosity, oil saturation and formation volume factor (reproduced with permission of Prof. Olivier Dubrule from his teaching slides).

This needs to be reconciled with the definitions of the Society of Petroleum Engineers (SPE) Petroleum Resource Management System (PRMS) designations of 1P (proved), 2P (proved plus probable) and 3P (proved plus probable plus possible) reserves. These are illustrated graphically in Figure 3.4, with qualitative descriptions of proved reserve given as having reasonable certainty.

The two-axis PRMS is illustrated in Figure 3.4. This a qualitative system based on descriptions but is derived from a quantitative probabilistic approach outlined above. Based on SPE nomenclature, 1P is therefore $0.13\,\text{mm}^3$, 2P is $0.26\,\text{mm}^3$ and 3P is $0.51\,\text{mm}^3$. This should be compared with the reporting of mineral resources outlined in Section 3.3.

4.6 Valuing Oil Companies

While there are obvious similarities in the resource and reserve nomenclature in petroleum and minerals, direct comparisons are probably not possible. The approach to extraction is of course also fundamentally different. Energy projects such as coal, uranium and oil sands are essentially mining operations although classified as petroleum projects. Here conventional mining techniques are used to produce oil, indicating that differences are becoming blurred. The investment community in any event often make little distinction between mining and petroleum — both sectors are commodity-based and are associated with products that have over the last decade been associated with unprecedented levels of volatility. The impact this has on revenue assumptions and determination of economic viability needs to be linked to concepts of probabilistic reservoir estimation techniques and simulation technology used to determine the underlying key relationship between "total petroleum-initially-in-place" and "estimated ultimate recovery".

Most oil and gas analysts understand that the 1P reserves (proven only), which the SEC require, are no basis for valuing the future earnings potential of a company. They will look at risk-weighted 2P (proved + probable) reserves. Furthermore the NPV of all oil and gas projects is based on a simple DCF using a standard discount rate of 10% (designated PV10) for SEC-reporting purposes.Companies are probably happy to report normalised NPV10 numbers to SEC. This convention is, however, rarely the basis of share price valuations.

Valuation of petroleum projects tends to be a niche skill that has been in demand by both the IOGs and the natural resource teams of investment banks. The current volatility with oil price is therefore likely to increase the career opportunities for professionals in the business development divisions of the natural resource groups as well as the financial services offering advice on restructuring.

4.7 Natural Gas and Oil Pricing

Natural gas is not priced globally because it cannot be easily transported, although the growing LNG business may change that somewhat in years to come. LNG is natural gas (predominantly methane, CH_4) that has been converted to liquid form for ease of storage or transport. It takes up about 1/600th the volume of natural gas in the gaseous state. Natural gas is condensed into a liquid at close to atmospheric pressure by cooling it to approximately −162°C. Maximum transport pressure is set at around 25 kPa (4 psi).

Butane is derived from a manufacturing process using natural gas as the feed and in liquid form is transported and traded globally when produced to an international specification. The hub of trade is Mt Belvieu, Texas, just outside Houston. This location serves as the swing for the global market, taking surplus into inventory (salt caverns) and exporting when needed to fill a deficit.

The petrochemical cracker operators are able to swing their feed from propane to "naptha" (light oil C_{5+} molecules, see Section 6.1) very quickly, and therefore they are extremely price-sensitive. Hence there is a well-known relationship between the Mt Belvieu reference price for propane and the price of light, sweet crude oil on the New York Mercantile Exchange (NYMEX), the commodity futures exchange and West Texas Intermediate (WTI) that naptha is referenced to. Cushing, Oklahoma is a major trading hub for crude oil and is the price settlement point for WTI which is also known as Texas light sweet and is used as a benchmark for oil pricing alongside the price of Brent crude from the North Sea.

Although there are short-term deviations from the relationship of Mt Belvieu reference price at WTI that can last for weeks or even a month or two, over the long term the volume/heat content relationship has remained consistent.

4.8 Real Option Valuation

During periods of low commodity prices more projects will be marginal but there may be the intuitive view that a project has intrinsic value. This can be determined quantitatively through the application of the technique of real option valuation (ROV). This is based on the Black–Scholes (BS) equation in which any deferred value for a marginal project can be estimated. The volatility variable used in the BS is derived from the distribution of the present value of free cash flows during the operational period based on a MC simulation of the relevant commodity price volatility. The time to expiry of the option would be linked to practical considerations such as the duration of an exploration licence. This in turn must be linked to the integrity of the award of licences and the principles of security of tenure once exploration results in a discovery and evaluation are completed. Conversion to a mining licence must be subject to compliance with environmental permitting.

This is the appropriate framework for the treatment of commodity prices and is discussed in more detail in Section 10.3.

If the application of ROV generates a value in the future which is significantly greater than the current value then it will be worthwhile retaining the asset. At the time of the expiry or at any stage up to its expiry (so this is an analogue of an American option not a European option which can only be exercised at the expiry date) you have retained the right but not the obligation to continue with the programme of evaluation or development. If the price of the commodity (gold, copper, oil, uranium, coal, etc) has increased than you proceed with further evaluation drilling or to a full technical feasibility study (FTFS) (see Section 8.5). If not you let the licence lapse and abandon the project.

All of the above follows on from the NPV optimisation process that is undertaken at the preliminary feasibility study (PFS) stage discussed in Section 8.3 and takes into account time-value of money, cut-off grades and cost-capacity relationships. This approach allows the promoter to identify the sweet spot for the project, but the NPV expressed as a deterministic outcome must then be considered from a probabilistic perspective. This then loops back to the role of MC simulation outlined above in the selected base case.

Mineral and Petroleum Geosciences

Chapter 5

Mineral Deposits

5.1 Introduction

The terminology that has developed around mineral deposits is a blend of historical use and more recent attempts to clarify nomenclature used in reporting mineral resources and ore reserves. The traditional connotation of an ore mineral was a museum-quality sulphide or oxide that was often geologically exotic in order to have been collected in the first place. Such specimens are seldom encountered in an active mining operation. The term "mineralisation", when used to describe a mineral deposit, has the connotation of a concentration of minerals able to support a commercially viable operation, but this use also needs to be treated with caution. The term grade is used to report concentrations of the metal or mineral and it is the discovery of mineralisation that combines high grade in association with a large mass of host rock that is the ultimate aim of an exploration programme. At that point a mineral deposit has been identified.

It is important to relate any description of a mineral deposit to the associated mineral or metal type, and this provides the basis for classification systems. Type deposits are described in great detail and analogues which have been assigned are identified. Inevitably overlap occurs, some deposits do appear to be unusual and in a class of their own, and all deposits will have their own unique geochemical signature and geological setting.

The field of mineral deposit studies is in turn inextricably tied to the discipline of earth science and the study of rocks. These comprise the three well-known groups. Igneous rocks fall into two broad compositional types: the high-magnesium and high-iron varieties (ultramafic and mafic rocks respectively) and high-silica and high-alumina varieties (salic rocks). These in turn will be classified according to their mode of emplacement — intrusive or extrusive. Sedimentary rocks are essentially the accumulation of eroded and re-deposited detritus of pre-existing rocks, of chemical precipitation or life processes. Metamorphic rocks are formed when igneous or sedimentary rocks have been subject to high pressure and temperature. These processes in turn are driven by the dynamic mechanism of plate tectonics. Mineral deposits can be located in any three of the rock groups while petroleum is restricted to sedimentary host rocks.

The geochemically abundant elements in the crust comprise silicon (27.2%), oxygen (45.2%), aluminium (8.0%), iron (5.8%), calcium (5.1%), magnesium (2.8%), sodium (2.3%), potassium (1.7%) and titanium (0.9%). The remaining elements comprise just 1% of the crust. These are the geochemically scarce elements. Unlike the petroleum sector, which is able to focus on hydrocarbons, the geochemically scarce elements include the metals that have practical industrial uses, which incorporate a significant portion of the periodic table. There is therefore a significant divide between geological models needed to understand the formation of mineral deposits and the minerals engineering needed to exploit them and the petroleum, geological and engineering skills needed to extract hydrocarbons. All are natural resources so the approach to investment funding is very similar.

Coal is a carbonaceous material also confined to sedimentary rocks and is covered in Chapter 6.

Each mineral deposit type will have its own unique geological characteristics and earth scientists and exploration geologists will usually specialise in a particular subset of a mineral deposit type. It is often prudent to bring in the relevant authority to provide an independent perspective of a deposit after discovery and before embarking on a detailed programme of evaluation and development of a project. It is all too easy to over-simplify the inherently complex physical and chemical processes involved in the formation of a mineral deposit. No matter how proficient the conventional engineering, or innovative the financial engineering, an operator cannot create a single additional ounce of precious metal or kilogramme of base metal that nature did not provide. This is why a basic understanding of mineral deposits is needed for anyone involved in the natural resource sector.

5.2 Lithogeochemistry

Chalcophile elements, according to the Goldschmidt classification, will partition strongly with the sulphur (S) and arsenic (As) anions and many of the transition elements that have important industrial uses fall into this category. These include nickel (Ni), copper (Cu), zinc (Zn) and lead (Pb). The platinum group elements (PGE) act both as siderphile and chalcophile metals and comprise platinum (Pt), palladium (Pd), rhodium (Rh), iridium (Ir), ruthenium (Ru) and osmium (Os). Platinum will combine with S and As but is also found as ferroplatinum in alluvial deposits. Gold of course is often present as native metal.

There is no shortage of the geochemically abundant industrial metals of Fe and Al or indeed the industrial minerals such as halite or limestone. To form high-grade ore quality deposits, metals or minerals need to be associated with high tonnages. While this is geologically rare, they will always be low-value commodities, which means that production will need to be associated with large capacity and the availability of infrastructure. For the geochemically scarce elements, when associated with a sulphide, oxide or present as an alloy or native metal, liberation must be achieved from the host rock matrix (also known as gangue). This permits separation into a concentrate and immediately raises the value of the product by weight (see Chapter 9).

A sulphide not only acts as a powerful collector of metals, once liberated from the matrix of gangue rock the technique of froth flotation ensures a highly efficient mechanism for its collection and concentration (see Section 9.6.1). Release of the metals from the primary mineral association also takes place as an exothermic reaction in the subsequent downstream pyrometallurgical process. Where metals partition into a silicate phase, release from the mineral

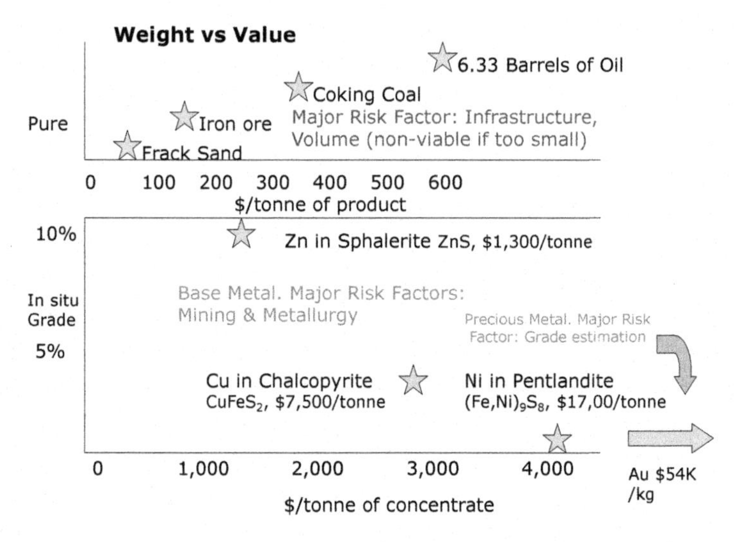

Figure 5.1. Relationship between value and tonnage. Top — geochemically abundant material. Bottom — geochemically scarce elements.

phase involves an endothermic reaction. This effectively represents an energy barrier that must be considered in any evaluation of a project.

The simple expression of the relationship between grade and metal price is illustrated in Figure 5.1.

5.3 Diversity of Mineral Deposits

A good example of the potential complexity that arises in assessing the economic potential of a mineral deposit hosting geochemically scarce elements, is that of the rare earth elements (REE). The most commercially important are the elements neodymium (Nd) and dysprosium (Dy) which in turn are just two of the 14 REE elements within the periodic table. Neod is used in the manufacture of powerful magnets. Dysprosium is present in control rods in nuclear reactors where it acts as a neutron absorber in nuclear fission reactions. It is also used in the manufacture of compact discs. Neither Nd nor Dy occur in nature as free elements but are found in the radioactive mineral monazite. This has the formula (Ce, La, Nd, Th)PO_4 containing up to 14% Nd. Dy is not a major constituent of monazite although it may be present in concentrations of up to 0.7% together with the other REE.

Differential separation of the REE from monazite into the pure form of each element requires sophisticated chemical processes.

Monazite is present as a mineral in a wide range of primary igneous rocks but given its relatively high hardness (5 on the Mohs scale) and density (5.15 g/cm^3) it will, when liberated by natural processes of weathering and erosion, concentrate in fluvial river, marine coastal and dune systems. As a consequence it is commonly associated with placer-type heavy mineral deposits which are accumulations of alluvium (gravels, sands and dunes). These in turn provide the feed for heavy mineral alluvial mining operations which produce, as a primary product, the titanium-bearing minerals ilmenite and rutile, and the zirconium mineral zircon.

In some placer-type heavy mineral deposits where monazite is present it may be produced as a by-product in the mineral separation process. To exploit a monazite resource these have to be associated with ilmenite, rutile and zirconium in concentrations and volumes sufficient to justify the investment in a heavy mineral alluvial mining operation. The economics of such an operation normally require a rate of production of about 10 million tonnes of feed per annum and an investment in mining equipment (earth moving and dredging) and processing equipment (gravity and electrostatic) of at least $50 million. The nature of the mining itself involves a major impact on the environment, creating significant obstacles in securing permits. The main areas of the world where monazite are produced are in the US, Brazil, India, Sri Lanka and Australia.

Outlined below are descriptions of some of the major mineral deposit types. There are notable omissions such as the Iron Oxide-Copper-Gold(-U-REE) deposits, "IOCG" types as well as the palaeoplacer deposits of the Witwatersrand in South Africa. A more comprehensive treatment of mineral deposits is given in an *Introduction to Ore-Forming Processes* by Laurence Robb (2004).

5.4 Volcanic-hosted (Volcanogenic Massive Sulphide)

The idea of massive sulphide formation by volcanic exhalative processes was proposed in the late 1950s for the sulphide ores associated with sedimentary lithologies of the Scandinavian Caledonides. Since then stratabound massive sulphides have been described throughout the world, ranging from the Archaean deposits of Canada to the currently active hydrothermal exhalations observed on the ocean floor ridge-rift systems (see Section 5.9). The term "volcanogenic massive sulphide" (VMS) is used to describe a whole range of deposits which, although they share some common features, may be highly variable in their age, mineralogy and associations. They are normally polymetallic and are associated with Cu, Zn, Pb, Au and Ag. They typically comprise one or more lenses of massive pyrite and chalcopyrite hosted by mafic volcanics and underlain by a well-developed pipe-shaped stockwork zone.

The Iberian Pyrite Belt has been regarded as the type area for VMS mineralisation. This description is based on Carol Halsall's PhD dissertation (1989) that considered the lithogeochemistry of the

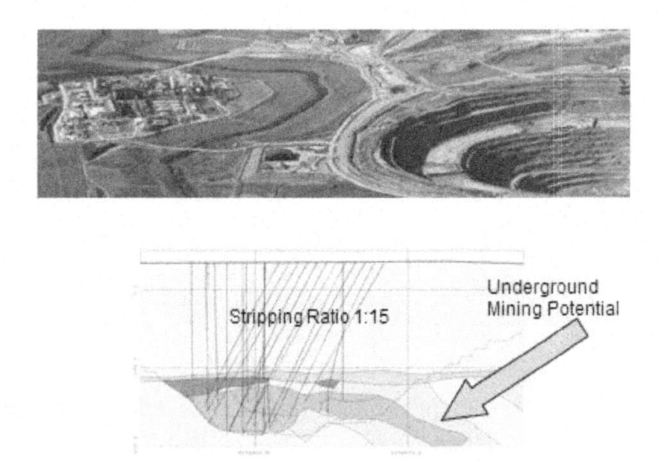

Figure 5.2. Las Cruces VMS deposit comprising 69,000 tonnes cu (2013 prodn. Operated by First Quantum Minerals). Reserves: 14.1 Mt @ 5.4% cu. Reproduced with permission from First Quantum.

Rio Tinto area. Although many of the mines have ceased production over the years, the discovery of the Las Cruces deposit below younger cover outside Seville has demonstrated that fresh discoveries remain to be made. Hydrothermal alteration of the footwall rocks is a characteristic feature of VMS deposits and a good understanding of the genetic process involved provides the foundation to successful exploration.

Conventional models for the genesis of VMS deposits by volcanic exhalation involve interaction between the host rock and large quantities of heated seawater, resulting in the leaching of metals from the footwall rocks and their subsequent deposition as sulphides on the sea floor. Although much of the metal transport and deposition in these systems undoubtedly takes place as a result of the convective leaching process, there is some controversy over the role of intrusive magmatisim in the generation of large massive sulphide deposits such as the Kuroko deposits of Japan.

5.5 Mafic and Ultramafic Magmatic Sulphide Hosted (Metal Group: Ni and Cu; Platinum Group Elements Pt, Pd, Rh, Ir, Ru, Os)

5.5.1 *Introduction*

Primary nickel sulphide deposits are hosted in ultramafic igneous rocks in Archaean greenstone terranes. Nickel grades are typically 1–5% and tonnages are moderate to high. Copper may be present as an economic by-product. Examples would be Mt Keith in Western Australia and Voisey Bay in Canada.

Orthomagmatic sulphide deposits are formed in large, stratiform, layered igneous complexes in Proterozoic cratonic settings. Their origin is intimately associated with the high-temperature processes of magma formation and crystallisation. Differences in the stages of geochemical evolution of the source magma results in the economic concentration of nickel and copper within certain layers, in addition to accessory PGEs. The Bushveld Igneous Complex of South Africa is a prime example, hosting the Merensky and UG-2 reefs. These were

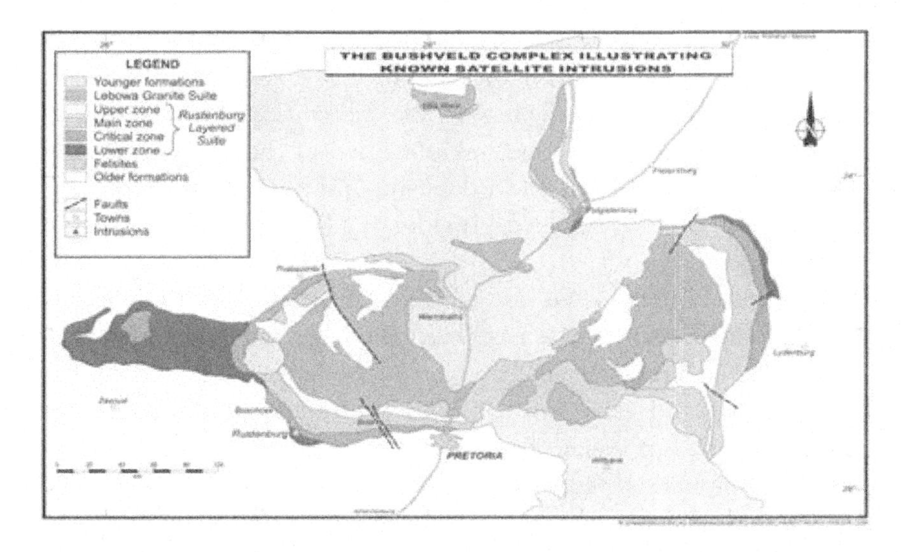

Figure 5.3. Geological map of the Bushveld Complex in South Africa. The Complex comprises five major subdivisions ("Zones") with a thickness of up to 7 km. It hosts 13 major chromite layers in the Critical Zone, major economic concentrations of PGE in the Merensky Reef, UG2 Reef and Platreef with vanadium being extracted from the Main Magnetite layer.

formed in closed chemical systems unlike the other major horizon, the Platreef.

5.5.2 *Fractionation*

When a large body of mafic magma is emplaced in the earth's crust, slow cooling takes place. Silicate, oxide and sulphide phases crystallise and sink through the magma chamber to form texturally distinctive layers. The removal of the more refractory minerals in this way depletes or enriches the residual melt in various elements through a process of fractional crystallisation. Enhanced levels of the elements nickel, chromium and vanadium are normally associated with the mafic rocks and enrichment during fractional crystallisation can be sufficient to form mineralised horizons at predictable levels within the intrusion.

Changes in melt composition during fractional crystallisation are illustrated in Figure 5.4. As can be seen, the Fe(Fe + Mg) ratio of

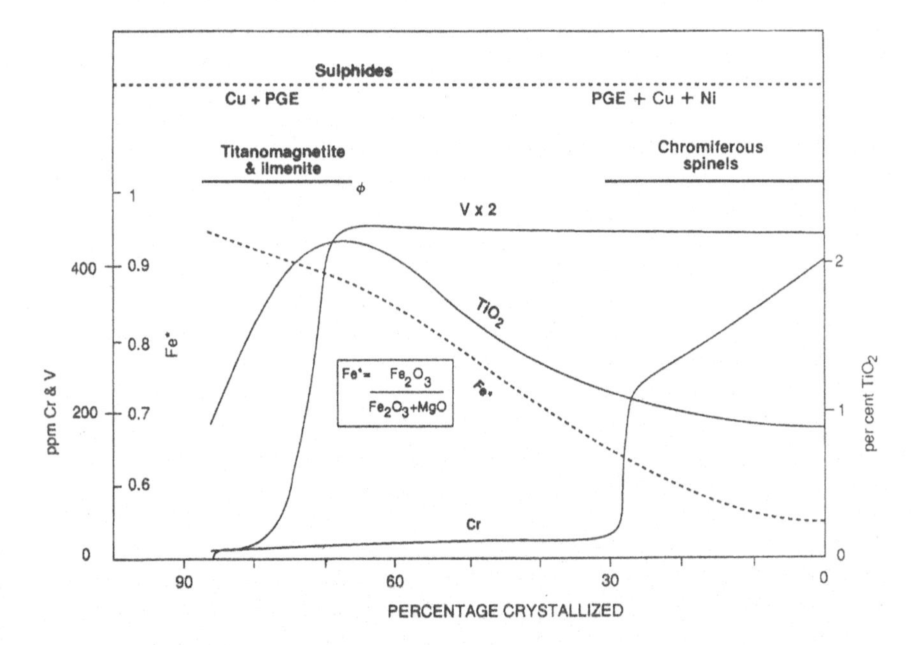

Figure 5.4. Model of fractional crystallisation for Karoo dolerites. Fe_2O_3 represents combined ferrous and ferric values.

the residual melt increases with fractionation. This is in response to crystallisation of magnesium-rich silicate phases such as olivine (Mg_2SiO_4) and pyroxene ($MgSiO_3$), which deplete the magma in magnesia.

Oxides are represented by ilmenite ($FeO{\cdot}TiO_2$) and the spinels. The latter represent a solid-solution series comprising the following end-members: spinel ($(MgO){\cdot}Al_2O_3$); magnetite ($FeO{\cdot}Fe_2O_3$); Magnesio–chromite ($MgO{\cdot}Cr_2O_3$); and chromite ($FeO{\cdot}Cr_2O_3$). See Section 5.9.1 for further detail.

Chromium levels in the melt decline in response to the early crystallisation of chromite. The proportion of chromium in the chromite declines with fractionation, while Fe ratios in the chromite are used as a broad index of fractionation. These are normally calculated from the mineral analyses given in weight % FeO, Fe_2O_3 and Cr_2O_3 by establishing the weight % metal from the molecular weight of the oxide. Cr/Fe ratios can also be calculated from stoichiometric atomic proportions. These apparently arcane concepts are central to any evaluation of a chromitite (a rock made up entirely of the mineral chromite) which is exploited for the production of ferrochromium.

Vanadium and titanium levels in the melt display a sharp drop in the later stages of the crystallisation history. Titanium depletion is due to the crystallisation of ilmenite. Vanadium partitions strongly into coexisting magnetite where V^{3+} substitutes for Fe^{3+} because of its similar charge and ionic radius.

5.5.3 *Sulphur solubility*

Studies undertaken on fresh submarine basalts, in which degassing has been inhibited by pressure of seawater, indicate that they contain significantly more sulphur that subaerially erupted basalts. The sulphur in the submarine basalt occurs as a dissolved constituent in the basaltic glass and as small immiscible sulphide droplets. These features can be reproduced in the laboratory. Figure 5.5 illustrates synthetically produced sulphide droplets consisting of an iron sulphide-oxide phase coexisting in equilibrium with a silicate melt containing 0.15 % sulphur.

Once formed, sulphide droplets in mafic and ultramafic magmas will act as collectors for copper, nickel, cobalt and PGE. The subsequent concentration of the metal droplets, normally through gravity settling, could lead to the formation of ore bodies. The tenor of the ore body will, however, be a function of the partitioning characteristics of the metals between silicate and sulphide melts, as well as the relative proportions of the melts. Early-formed sulphide melts will be rich in nickel and on cooling will crystallise pyrrhotite ($Fe_{(1-x)}S$) and pentlandite ($(Fe, Ni)_9S_8$).

Sulphides in most intrusive mafic rocks are evenly distributed as a finely disseminated phase. This suggests that the sulphide phase nucleated only at a late stage in the crystallisation history of the magma, allowing insufficient time for settling. The coexisting residual melt will also have been depleted in many of the economically important metals because of their preferential partitioning into earlier crystallising silicate and oxide phases. In the absence of a sulphide phase, nickel will partition into olivine while the silicate melt will become enriched in copper and PGE. The ratio of Cu/Ni increases, therefore, with fractionation. Subsequent precipitation of a sulphide phase will result in the crystallisation of pyrrhotite, pyrite and chalcopyrite ($CuFeS_2$).

For a mafic magmatic sulphide ore deposit to develop, it is clearly important that the sulphide phase should form very early in the crystallisation history. This would probably require an open system in which external factors can influence the timing of sulphide immiscibility. The characteristics of these factors (based on studying magmatic ore deposits in natural rocks) is limited by the difficulty of obtaining reliable data for their intensive parameters such as temperature and density at the time of formation. By examining experimental systems at magmatic temperatures under controlled conditions it is, however, possible to model conditions required to initiate the precipitation of an immiscible sulphide phase. The results of theoretical and experimental work on sulphur solubility systems undertaken as

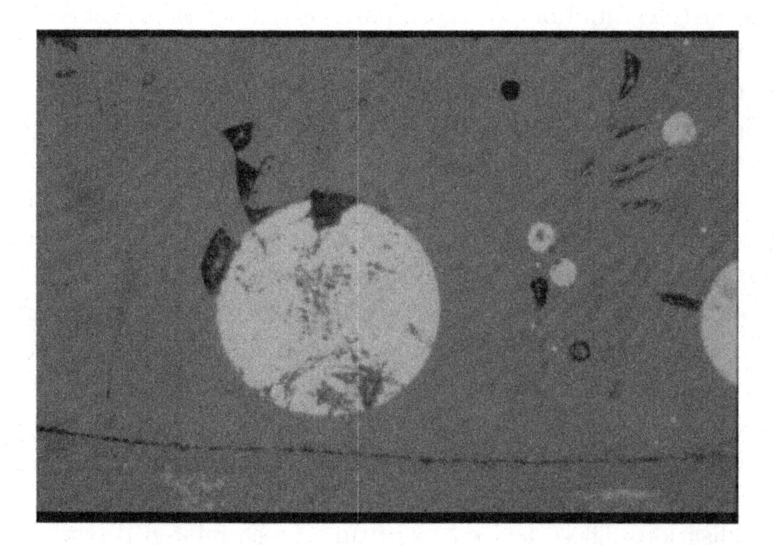

Figure 5.5. Reflected light micrograph of an experimental charge showing immiscible sulphide droplets in a silicate glass. Horizontal field of view = 40 microns. This sulphide-silicate immiscibility is equivalent to the metallurgical matte-slag system.

part of a PhD and postdoctoral year at Imperial College is given in Buchanan (1988).

The results of the experimental modelling of silicate-sulphide systems at constant temperature demonstrate that the sulphur-carrying capacity of a silicate melt can be increased by adding iron or a reductant. Clearly a magma contaminated with banded ironstone and organic matter would have a high sulphur-carrying capacity. The possibility of an immiscible sulphide forming in these circumstances is further enhanced where additional sulphur is introduced into a mafic magma such as the Bushveld-type.

Although the Platreef represents the product of primary crystallisation from an original Bushveld magma, contamination from floor rocks played a significant role in its petrogenesis. The horizon therefore provides a superb natural laboratory for testing and applying the results of the experimental work. The author mapped a 25 km stretch of the Platreef outcrop in the early 1980s, which provided the basis for developing the system of zoning for the Platreef that is still used. The author was then involved with the first geostatistical resource estimations of those parts of the Platreef where Anglo America Platinum established their Mogalakwena mine, currently the world's largest open-pit PGE mining operation.

High background levels of PGE of between one and three grams per tonne are present along most of the 25 km strike length of the Platreef, but the effect of contamination often obscures the primary trend of mineralisation. Although there are sectors along the Platreef where background levels of PGE rise to significant concentrations with grades of up to 56 grams/tonne recorded in the Zwartfontein South area, the key to delineating areas of mineralisation that can be successfully mined lies in distinguishing between erratic value and that which displays some degree of geological continuity. The task is aided by the recognition of several petrographic facies along the strike length of the Platreef, each with its characteristic style of PGE mineralisation. These facies can in turn be correlated directly with changes in floor rock geology.

The case history of the Platreef demonstrates the synergies that arise from esoteric scientific research when applied to mineral deposits. The principles were applied to other geological settings where similar conditions are present, notably the Nkomati mine, also in South Africa (see Figure 5.3).

5.6 Hydrothermal Precious Metals (Au and Ag) and Base Metals (Cu–Mo–Au)

5.6.1 *Epithermal gold*

It is generally accepted that the bulk of the world's primary gold deposits were formed through the action of hot aqueous (hydrothermal) fluids in the Earth's crust. Typical hydrothermal fluids responsible for base and precious metal deposits are Na-K-Ca-Cl-SO_4 brines, carbonated CO_2-H_2O fluids or mixtures of both. Ore metal concentrations are strongly dependent on the fluid composition, temperature and pressure, but are generally of the order of a few parts per million (ppm) to several hundred ppm for base metals and a few ppm to a few parts per billion for gold. Precipitation of the base and precious metals may take place in response to changes in the composition of the fluid either by mixing of two different fluids, by exsolution of a vapour phase (boiling) or by chemical interaction with the country rocks. Precipitation can also be caused by a drop in temperature or pressure.

Hydrothermal gold deposits were originally classified according to the likely source of fluid, the temperature and depth of emplacement — they include hypothermal, mesothermal and epithermal. Of these, epithermal has found a niche in the scientific literature and refers to deposits formed close to the surface and usually of tertiary age, and are the most important in terms of tonnage and contained gold. The type area is Carlin in Nevada.

5.6.2 *Porphyry copper*

Porphyry deposits are normally large tonnage (20 to more than 500 Mt) and low grade (0.4%–0.8% contained copper). They are amenable to bulk mining techniques (e.g. open pit, block caving). These are usually disseminated deposits which are typically associated with tertiary or recent, high level, acid to intermediate, subvolcanic intrustions. An important metallogenic province is the Cordilleras of South and North America. The type deposit is Bingham Canyon in Utah. The magmas are derived from the partial melting of a predominantly igneous source and of deep crustal or mantle origin. The host rocks comprise tonalites (quartz diorite) to granodiorite and tend to be metaluminous ($Al_2O_3 < Na_2O + K_2O$). They are relatively oxidised ($Fe_2O_3/FeO \geq 0.3$) and water-poor (2–3 wt.%). Magma can reach shallow depths (2–5 km) before crystallisation and water saturation will occur. They tend to occur on the oceanic side of a subduction zone.

Porphyry deposits comprise both hypogene mineralisation consisting of a stockwork and disseminated mineralisation (formed beneath the surface at high temperatures ~600–350°C). They can also be associated with secondary supergene and hypogene mineralisation enrichment due to acidified groundwater leaching which forms a surface blanket.

5.6.3 *Archaean greenstone gold deposits*

These deposits comprise structurally controlled vein, lode and replacement bodies associated with shear zones within Archaean greenstone terranes. These belts are composed of arcuate, greenschist-facies metamorphic volcanic or volano-sedimentary units in association with large areas of granite. Notable examples are in Western Australia.

Individual deposits are small, of the order of a few million tonnes of ore. The grades are usually higher than those in other types of gold deposits. Gold is typically associated with iron sulphides. Intense localised hydroythermal alteration of the wallrocks adjacent to the mineralised lodes is typical.

5.7 Granitic-hosted (Sn and W) and Nb–Ta

Host rocks to Sn and W mineralisation tend to be derived from the partial melting of a predominantly reduced sedimentary source with mid- to deep-crustal origin. These are often true granites, peraluminous ($Al_2O_3 > Na_2O + K_2O$) and reduced ($Fe_2O_3/FeO < 0.3$). They are water-rich (7–8 wt.%) and typically crystallise in the mid- to lower-crust (≈12–40 km depth). They tend to occur on the continental side of a subduction zone.

Pegmatites, coarse-grained zones within granites, also host the mineral "coltan" which consists of an intergrowth of minerals within the columbite-tantalite group (Fe, Mn)(Nb, Ta)$_2$O$_6$ for which a nearly complete solid-solution series exists. Tantalum (Ta) is widely used in mobile phones. A pure tantalite end-member could theoretically contain up to 86% Ta_2O_5. With specific gravities varying from 5.2 for columbite to 8.0 for tantalite, the minerals are often present in alluvial deposits. Given that tantalite can reach a price of over $200/tonne this encourages exploitation by artisanal miners in central Africa that is fuelling conflict.

5.8 Sedimentary-hosted (SEDEX and Carbonate) Pb, Zn, Ba

Sedimentary-hosted Pb/Zn deposits are associated with sedimentary exhalative Pb–Zn (SEDEX) and carbonated-hosted deposits. The latter were first described in the scientific literature as Mississippi Valley-type but the style of mineralisation is also present in the Irish-type deposits. Mineralisation mainly occurred sub-seafloor by replacement of carbonates during diagenesis. The source of the mineralising fluids is considered to be evaporated seawater and the main depositional mechanism is considered to be the mixing of metalliferous ore fluid with low-temperature bacteriogenic H_2S-rich brines.

The term stratiform Zn–Pb–Ba SEDEX deposits was originally introduced as a genetic term and refers to the ore-forming process that is found in local basins or depressions on the seafloor. Examples are: Sullivan and Howard's Pass (Canada); HYC or McArthur River and Mt Isa (Australia); and Red Dog (Alaska, USA). They contain more than 50% of the world's reserves of Pb and Zn and more than 25% of global production. The dominant sulphides are galena, sphalerite, pyrite and barite, with lesser amounts of chalcopyrite. Silver can be present in minor amounts (mostly in arsenic and antimony sulphosalts or in galena). Ore minerals commonly occur in

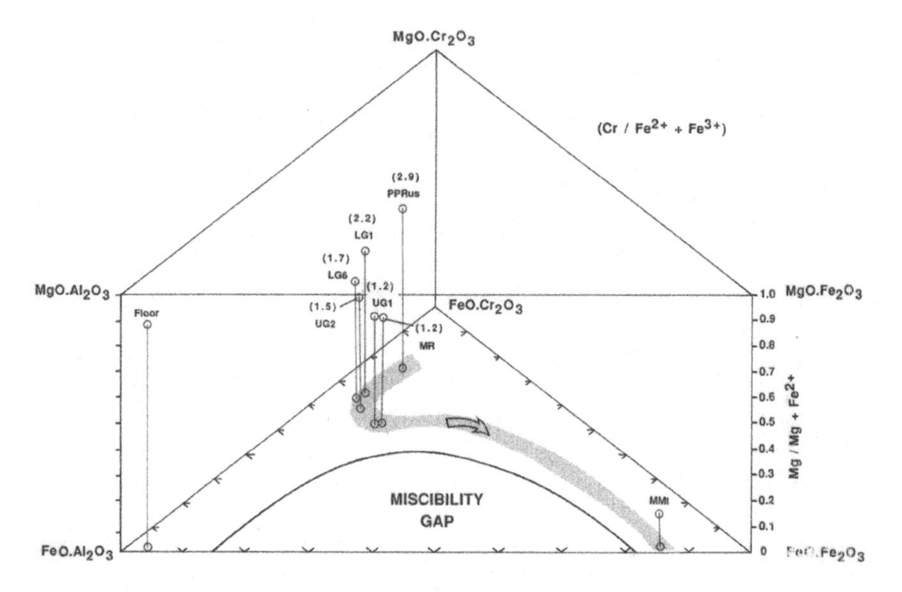

Figure 5.6. Compositional variations of spinel phases from the Bushveld Complex plotted within the spinel prism.

finely-layered horizons, parallel to bedding, interbanded with clastic material and iron sulphides.

5.9 Geochemically Abundant Elements Cr, Fe

5.9.1 *Spinels in mafic magmatic rocks*

Ferrochromium and ferrovanadium production in South Africa is based on exploiting the chromite and magnetite deposits of the layered rocks of the Bushveld Complex. The compositional field of spinels defined by their atomic end-members is given in Figure 5.6. The use of Cr/Fe ratios to describe compositional characteristics of the spinels can be misleading, as samples with the same chromium levels would have different molecular Cr/Fe ratios where the proportion of MgO.FeO varied. Accurate comparison of the composition of different samples should also take into account atomic solid solution between the divalent (Fe^{2+} and Mg^{2+}) and trivalent (Fe^{3+}, Al^{3+} and Cr^{3+}) elements.

A solid solution can exist theoretically between chromite and magnetite, producing a continuous fractionation trend. The minerals spinel and magnetite will coexist as separate phases. A spinel phase is not, however, normally ubiquitous throughout a fractionated sequence of rocks. In the Bushveld Complex the spinel is absent in the Main Zone rocks, which overlie the Merensky reef. The reappearance of a spinel phase as magnetite is observed in the Upper Zone, 2,500 m higher in the sequence.

Numerous chromitite layers are present in the 1,500 m Critical Zone of the Bushveld Complex, which underlies the Merensky Reef. They are categorised, from the base upwards, into Lower Group (LG), Middle Group (MG) and Upper Group (UG). Chromite operations, which supply ore for the local ferrochromium industry, are concentrated in the LG layers, where the highest Cr/Fe ratios are found. Normally only the most persistent and thickest layer in the group is mined. In the western and eastern parts of the Complex this is the LG-6, which is up to 2 m thick and has a Cr/Fe ratio of 1.6.

The UG of chromitite layers is characterised by Cr/Fe ratios of 1.4 or less. The UG-2 would not be exploited for its chromitite alone, but it does yield chromite as a by-product of platinum mining, which is not normally utilised. The UG-1 occurs 20 m below the UG-2 and consists of a composite zone of alternating narrow chromitite and anorthosite layers and is exposed in a classic location at Dwarsrivier in the eastern limb of the Bushveld Complex (see Figure 5.7). There have been many attempts to create a business case around the premise that as UG-2 is a by-product of platinum mining no cost can be assigned to extraction. The product needs pelletising and as ferrochromium producers are paid on chromium units, the best that has been achieved is to blend UG-2 with LG-6.

The LG-6 chromite produces charge ferrochromium with 54% chromium. The ferrochromium producers derive their revenue from chromium units in ferrochromium. As smelting costs remain the same, chromite with high Cr/Fe ratios command higher prices. Ideally, lumpy chromite ore should be used to produce ferrochromium as it produces a permeable charge.

Figure 5.7. View of the Dwarsrivier cutting in South Africa showing the UG-1 layers.

Magnetite and ilmenite ($FeTiO_3$) are present as a discrete phase in nearly all the Upper Zone rocks of the Bushveld Complex, up to 21 individual layers occurring in the eastern limb. The thickest of these is the main magnetite layer, which is up to 2 m in width. Since vanadium partitions strongly into the first magnetite to crystallise and the magma becomes depleted in vanadium as crystallisation proceeds, the main magnetite at the base of the Upper Zone carries the highest vanadium values (up to 1.6% V_2O_5). These high levels, combined with its considerable thickness, represent a major resource of easily exploitable vanadium ore for the South African ferrovanadium industry.

The presence of Ti as illustrated in Figure 5.8 means that magnetite associated with mafic magmatic layered deposits cannot be used as a primary source of iron ore.

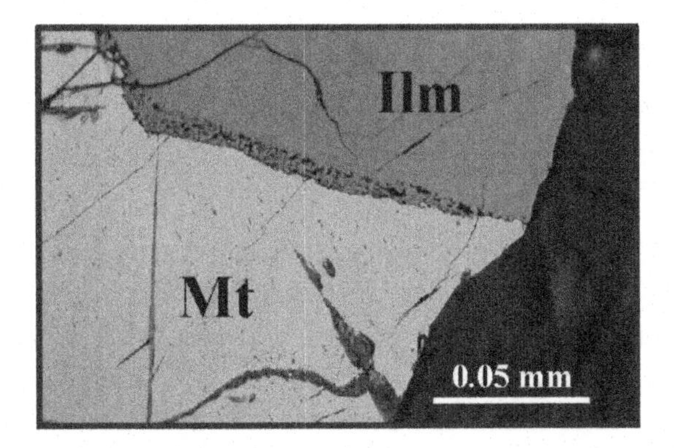

Figure 5.8. Reflected light micrograph from a Bushveld magnetite layer demonstrating the intergrowth of magnetite (Mt) and ilmenite (Ilm). The titanium mineral ulvospinel has exsolved from the magnetite, which means that while magnetite can be separated from the ilmenite, titanium would remain associated with magnetite.

5.9.2 *Iron formations*

Iron ore is associated with both chemical sedimentary-banded deposits where $Fe2^{+}$ was transported in solution and chemically precipitated via oxidation to $Fe3^{+}$ in hematite Fe_2O_3 or magnetite $FeO{\cdot}Fe_2O_3$. Iron is also present in skarn deposits as well as carbonatite.

The Fe content of hematite is 69.94% and it has a density of 5. It is associated with banded iron formations (BIFs) and accounts for most of the world's production of Fe.

These are very large deposits in stratigraphic units hundreds of metres thick, hundreds to thousands of kilometres in lateral extent. They are characterised by fine layering, typically 0.5–3 cm thick alternating dark Fe-rich (most commonly hematite, more rarely magnetite) and light, silica-rich (chert) bands. Most BIFs were deposited in the Archean and Palaeoproterozoic between 3500–1500 Ma. Silica (SiO_2) precipitated from dissolved $Si(OH)_4$ in seawater and therefore there is a lack of associated aluminous and siliceous clastic particles. Where BIF undergoes secondary enrichment then ore grade iron deposits are formed via supergene and/or hydrothermal processes grading 62–65% Fe. This is known as "direct shipping ore" because no processing is required, but see Section 8.7.2 for discussion on grade optimisation.

5.10 Marine Minerals

The seabed is host to a range of minerals, including the diamondiferous gravels found off the west coast of South Africa and Namibia. "Marine minerals" in the context of this report relate to "polymetallic nodules" under the International Seabed Authority (ISA) nomenclature and are more commonly described in the scientific literature as "manganese nodules". These are not restricted to the Clarion/Clipperton Zone, where all the contracts are held. The recognition of "ferromanganese crust" associated with seamounts appears to represent a new class of deposit, but this appears geochemically to be very similar to conventional manganese nodules.

"Polymetallic sulphides" under the ISA nomenclature are more commonly recognised as "seabed massive sulphides". The most advanced of all the proposed marine mining projects is the Nautilus Minerals *Solwara 1* project, which lies in the territorial waters of Papua New Guinea and aims to exploit polymetallic sulphides.

Chapter 6

Petroleum and Coal

6.1 Introduction

In organic chemistry a hydrocarbon is a compound consisting entirely of carbon and hydrogen. The major constituents of petroleum and coal are hydrocarbons and both are derived from life forms (living organisms). These are present as alkanes and aromatics with the general formulae C_nH_{2n+2} and C_nH_{2n-6} respectively. The major gaseous hydrocarbons comprise CH_4 methane (also described as natural gas), C_2H_6 (ethane), C_3H_8 (propane) and C_4H_{10} (butane). Petroleum refers to hydrocarbons from C_5 to C_{17} that are in a liquid state. Hydrocarbons from C_{18} to C_{40} make up waxes.

When the gas is removed from the reservoir during production, it cools and pressure is lowered: some of the C_{5+} molecules condense out as a liquid phase known as condensate which is a valuable product in itself. It is essentially a light crude oil.

Clathrates have water-based molecular structures that capture methane under appropriate high pressure and low-temperature conditions, and are common in permafrost environments.

The coalification process of precursor accumulated plant material is illustrated in Figure 6.1. Initially the main influence is biochemical activity followed by burial and then subsequently, thermal and confining pressure. When combined with geological time anthracitic coal is formed which comprises mainly amorphous carbon. Coal has complex physio–chemical characteristics that need to be determined when considering end use. This includes a range of different qualities such as feed for electrical power generation, injection of pulverised coal into a blast furnace during steel production and conversion into coke.

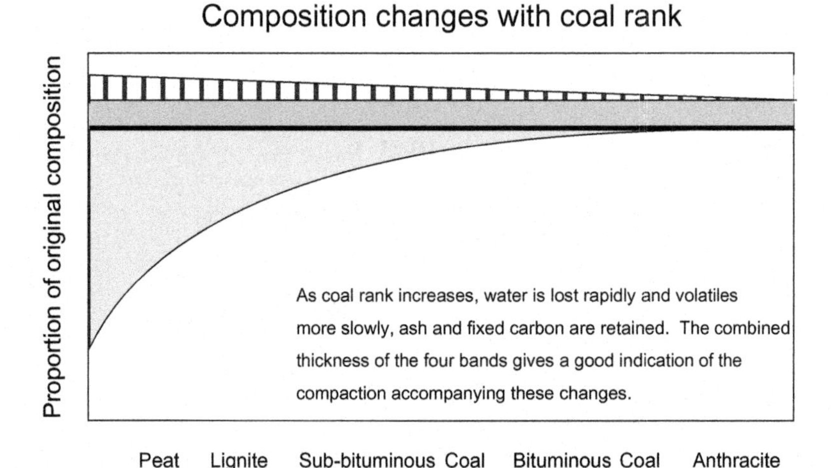

Figure 6.1. Diagram demonstrating the transition from transformation of vegetable matter in peat though to coal.

Methane is an important constituent of coal and while representing a significant hazard in an underground coal-mining operation, can be extracted ahead of mining by drilling wells into the coal seam. Through different applications of borehole technology, methane is extracted from coal seams both for the purpose of providing safe underground mining conditions in coal mines, and as the source of unconventional natural gas as a cleaner fossil fuel energy. Pressure is reduced by dewatering and the desorbed methane flows to the surface, compressed and used as an energy source. Unmined coal beds are also targets for carbon dioxide sequestration as this can be used to enhance coal-bed methane recovery.

The precursor to petroleum formation is considered to be preserved high molecular weight organic matter from dead organisms that form kerogen in an evolving sedimentary basin where anoxic conditions are present during initial deposition. Because of its high molecular weight, kerogen is normally insoluble in organic solvents. Type-I kerogen forms mainly from algae, Type-II kerogen from a combination of marine and plant organic matter and Type-III kerogen from land plants.

Clearly for hydrocarbons to accumulate, a source rock is needed. As with the coalification process, moderate temperature combined with pressure is required. Hydrocarbons need to be heated in the range 80–150°C under lithostatic pressure (the oil window). Over geological time this will result in the precursor complex organic chemical molecules being broken down as a consequence of cracking into simpler molecules. At lower temperatures conversion is incomplete and at higher temperatures pure methane is formed up to the phase boundary with graphite.

Type-I and Type-II kerogen are considered to be the more likely sources of petroleum while Type-III kerogen is more likely to produce gaseous components, mainly methane. Under more extreme pressure and temperature conditions hydrocarbons derived from Type-I and Type-II kerogen will also revert to gaseous components. Coal is formed from preserved plant material and is classified as Type-III kerogen.

Chemical engineering through the Fischer–Tropsch (F–T) process is able to reverse the natural coalification processes and convert a feedstock of coal, when reacted with steam and methane, into constituent alkanes and aromatics to produce synthetic fuels and a range of complex organic chemicals. Sasol in South Africa operates the Secunda coal-based synfuels manufacturing facility which converts mixtures of carbon monoxide derived from coal and hydrogen into liquid hydrocarbons. Sasol are actively seeking to promote their technology in which development of gas resources will enhance the gas to liquid capabilities in Sasol's core F–T process, while seeking to utilise carbon dioxide generated to recover residual hydrocarbons from mature oil reservoirs. There is also an angle on securing carbon credits.

Any consideration of energy derived from hydrocarbons needs to recognise that the chemical systems, whether formed in a geological setting or in the controlled environment of an industrial process, are therefore constrained by the same basic thermodynamic constrains.

6.2 Wessex Basin Petroleum System

The type area for a petroleum system is located in southern England where the full sequence of sediments from the Permian through to the Cretaceous (see Figure 6.2) can be seen in coastal exposures. The sediments dip to the east so successively younger rocks are encountered in a traverse from Exeter in the west through to Kimmeridge Bay in the east. The basin contains the largest onshore accumulation of oil in north-west Europe at Wytch Farm with a STOIIP (see Section 4.5) of 800 million barrels (MM bbls) of oil and a total recoverable reserve of about 450 MM bbls. Directional drilling from the site extends up to 30 km off-shore.

The Permo-Triassic rocks were deposited in continental desert environments and are shown in Figures 6.3 and 6.4. Both the Exmouth sandstone and the Sherwood sandstone make excellent reservoir rocks due to their high permeability and porosity. The former is the reservoir rock for the southern North Sea oil field and the latter for the Wytch Farm oil field.

Porosity, usually abbreviated to Φ, is a measure of the storage capacity of the reservoir. It is defined as the ratio between pore volume, Vp, and bulk volume, Vb, and expressed as a percentage:

$$\frac{\mathrm{Vp}}{\mathrm{Vb}} \times 100.$$

The most common porosity range is 10–20%; the highest recorded porosity value is 37% and the maximum theoretical porosity value is 47%. Permeability measures the ability of fluids to flow through rock (or other porous media). The darcy is defined using Darcy's law and typical values of permeability range as high as 100,000 darcies for gravel, to less than 0.01 microdarcies (mD) for granite. Sand has a permeability of approximately 1 mD.

An example of a potential reservoir rock is present at Orcombe Rocks shown in Figure 6.3. The rock is poorly cemented with high porosity and a permeability of around 100 mD. This unit is the equivalent of the Rotlegend reservoir in the North Sea oil field.

A further example of a reservoir rock is present at Budleigh Salterton and is shown in Figure 6.4. Here the Sherwood sandstone is located at the top of the Permian and base of the Triassic. High porosity of 15–20% and permeability in the range 10–100 mD is present. This unit is the main reservoir for the Wytch Farm oil field with a STOIIP of 300 MM bbls.

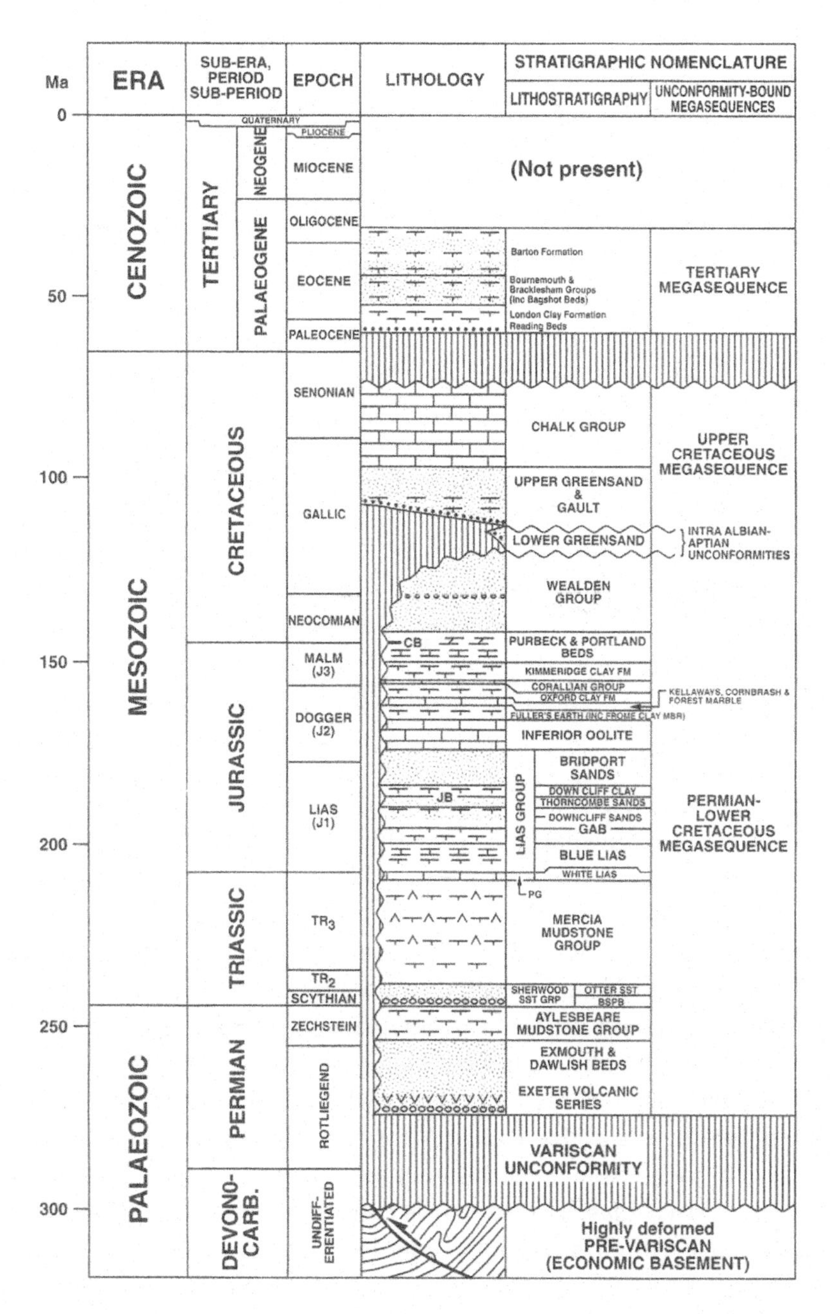

Figure 6.2. Stratigraphy of the Wessex Basin taken from Underhill, J.R. and Stoneley, R. 1998.

Figure 6.3. Orcombe Rocks. View of large-scale Aeolian cross-bedding in Permian-aged Exmouth sandstone comprising medium-sand sized rounded grains.

For a petroleum system to form an impermeable cap, rock needs to be present immediately above the reservoir rocks. The Mercia mudstone provides this feature for the Wytch Farm oil field and the contact is shown in Figure 6.5 Note that the transition between Sherwood and Mercia is gradational over tens of metres — at the base of the cliff is a thick sandstone with thin mudstone, above that is thick mudstone with very little sandstone present. There is uncertainty as to the position of the boundary which would have implications for defining the top of a reservoir. Permeability in the Mercia is 10–20 mD, providing a suitable cap rock for the Wytch Farm oil field.

Clearly for petroleum to accumulate there must be a source and the organic-rich Lias (Figure 6.6) is considered to be a candidate for the Wytch Farm oil field, but in the North Sea burial was too deep and both temperatures and pressures were outside the oil window. Implicit in this model is the requirement for the hydrocarbons to migrate from the source rock into a suitable reservoir with a seal.

Figure 6.4. Budleigh Salterton. View of interbedded matrix-supported conglomerates and sandstones at the base of the Sherwood sandstones group with channel structures present, indicative of a fluvial braided river environment.

Three rock types are present at Charmouth: Limestone-marl-black shale. It combines characteristics of both marine and fresh water depositional environments but is predominantly marine and highly fossiliferous. The black shales contain between 3% and 4% total organic content and is probably the source rock for hydrocarbons for the Wytch Farm oil field.

The heterogeneity that can be present in a reservoir is best illustrated by the exposure of the Bridport sandstone at West Bay (Figure 6.7). Differential calcite cementation forms harder beds more resistant to erosion. These discontinuous layers have a significant impact on fluid movements in the subsurface. Porosity in softer beds is 10–100 mD and much lower in cemented beds. The Bridport sandstones are the reservoir rocks for the first discovery at the Wytch Farm oil field before the larger underlying Sherwood sandstone-hosted reservoir was discovered.

Figure 6.5. Sidmouth. View showing the top of the Sherwood sandstone (see Figure 6.4) and the transitional contact with the overlying Mercia mudstone.

This was the first reservoir to be discovered at Wytch Farm and contains 150 MM bbls of STOIIP, smaller than the underlying Sherwood sandstone-hosted reservoir. As would be expected, lateral migration of oil would be relatively unimpeded but although the harder resistant beds are persistent, vertical migration would be restricted. As a consequence recovery from the reservoir is about 20%.These sedimentary features would not be imaged by a seismic survey and even if a well log did identify the presence of bedding with different hardness characteristics, the lateral characteristics would be a matter of speculation. There is an inherent uncertainty associated with this type of reservoir that should be considered at the investment stage. The heterogeneity would also influence the performance of the reservoir during production as water injection aimed at sweeping oil towards a well system would not retain a simple interface but fingering would take place and pockets of oil would be isolated.

Figure 6.6. Charmouth. View showing the Lias, a finely-bedded succession that includes some organic rich mudstone located at the base of the Jurassic.

Migration of oil from a source rock can be seen at the Osmington Mills exposure (See Figure 6.8). The well sorted sandstone is interbedded with bioclastic limestones and has a porosity of 200 mD. Heavy oil is seeping from the beds at the base of the cliff, suggesting that this is an exhumed reservoir. The original oil has probably been biodegraded.

The Wessex petroleum system reflects the dynamic tectonism that started with the Hercynian orogeny and resulted in the local north-south stress regime that produced the graben structures that permitted basin development and the deposition of the Permian and Jurassic rocks. The subsequent Alpine orogeny which started at the end of the Cretaceous and remains active produced north-south compression and resulted in the folding of the earlier basin sediments. This folding and faulting created the spatial relationships that permitted hydrocarbon-source rocks such as the Lias to enter

Figure 6.7. West Bay. View of Bridport sandstone which lies above the Lias in the Jurassic, comprising fine- to medium-grained sandstone suggested to be shallow beach deposits.

the petroleum window, and oil to migrate and then to accumulate in suitable structural traps. The impact of the Alpine deformation can be seen from the Lulworth crumple (Figure 6.9). The deformation has been caused by the Alpine orogeny and reflects the dynamic tectonism present in the Wessex Basin.

Tectonism also provides the mechanism for reservoirs to become over-pressured as oil trapped in suitable structures is then subject to lithostatic pressure.

Burial of sediments during basin development will also result in compaction of sediments and migration of both water and hydrocarbons.

Chalk is also present in the Wessex basin (see Figure 6.10). This rock is normally characterised by micro-porosity but is the main reservoir rock in the EkoFisk oil field in the southern and eastern North Sea oil field where interconnected fractures result in high permeability.

Figure 6.8. Osmington Mills. View of the Bencliff grit which is part of the Corallian Group present towards the top of the Jurassic showing an oil seep.

A view of Kimmeridge Bay and the production facility is given in Figure 6.11. At its peak the well produced 100 bbls per day and was originally estimated to have a STOIIP reserve of 1 MM bbls. Actual recovery has been some 3 MM bbls, suggesting that oil is leaking from a breached trap, or hydrocarbon is still being generated in an underlying source or the original estimate of the size of the reservoir was incorrect. It is currently producing 15 bbls per day.

6.3 Unconventional Hydrocarbons

Unconventional hydrocarbons consist of unconventional petroleum, which are essentially alkanes with chain length up to C30+ and unconventional gas which is methane. These include tar sands, shale oil and oil shale derived from Type-I and Type-II kerogen. Unconventional gas includes shale gas which, although mainly methane, is

Figure 6.9. Lulworth crumple, Stair Hole. View of the folded and faulted alternating hard and soft Purbeck beds overlying the Portland limestone of the Upper Jurassic dipping at 45° to the north.

also derived from Type-I and Type-II kerogen. Coal-bed methane can also be considered as an unconventional gas, albeit derived from carbonaceous shales formed from plants not from marine organisms. The main product is methane, chemically indistinguishable from coal-bed methane and formed in the same terrestrial geological environment as coal. Coal-bed methane is an important source of gas from the southern North Sea. While the petroleum industry is considered as a separate business from coal, geologically and technically there are more similarities than differences. Both are essentially industries exploiting fossil fuels that provide most of the world's energy needs.

The characteristics of unconventional hydrocarbons are given in Table 6.1.

The idea that hydrocarbons are produced by heating sedimentary organic matter was developed from the technique of pyrolysis. This is thermochemical decomposition of organic material at elevated temperatures in the absence of oxygen which produces a petroleum-like product. This industrial process was undertaken at Kimmeridge Bay

Figure 6.10. Lulworth Cove. View of an outcrop of the chalk present at the base of the Cretaceous and formed from the accumulation of coccoliths, the individual plates of calcium carbonate formed by algae.

on the organic-rich sediments (oil shales) (see Figure 6.12). The oil shales were extracted as a mining operation worked on various occasions from the 17th century to the 19th century. While the sediments did not reach the oil window in the type location they are the source of petroleum in the North Sea oil field. At the type location it is classified as an oil shale and when pyrolysed will produce oil and was mined at the site for this purpose. This is an example of an unconventional petroleum resource. Heating oil shales in a hydrous

Figure 6.11. Kimmeridge Bay. View of pumpjack on the apex of an anticline. Seismic target drilled in the 1970s below the Kimmeridge clay that acts as the seal intersected the Cornbrash reservoir.

environment generates products that are physically and chemically closer to natural petroleum products.

The Athabasca tar sands (also called oil sands) are large deposits of bitumen or extremely heavy crude oil, located in north-eastern Alberta, Canada, centred on the town of Fort McMurray. These oil sands, hosted primarily in the McMurray Formation, consist of a mixture of crude bitumen (a semi-solid form of crude oil), silica sand, clay minerals and water. Where they are near to the surface extraction is undertaken as an open-pit mining operation. As will be noted, there is convergence between the petroleum and mining industries at the extraction stage.

When buried at depths where open pit mining is not practical then steam-assisted gravity drainage is used to extract the hydrocarbon. The heavy bitumen then has to be upgraded to light sweet and

Table 6.1. Classification of unconventional hydrocarbons.

Type	Geological setting and rock type	Hydrocarbon characteristics and reservoir type	Type area and company	Extraction and processing method
Unconventional Petroleum				
Tar Sands	Shallow exhumed conventional reservoir in which light fraction has been lost.	Viscous bitumen present in sand and clay.	Alberta, Canada	Open pit mining or *in situ* production involves injecting super-heated steam. Energy intensive. Syncrude
Shale Oil	Deep sea (anoxic) marine environment. Organic-rich black shales that have reached the "oil window" pressure and temperature field of 60°C to 80°C	Liquid oil present trapped in pores of the shale. Some oil may migrate into a conventional porous sandstone reservoir.	Bakken and Barnett shales of North America	Tight oil requiring horizontal drilling and hydraulic fracturing (fracking). Water, sand and chemicals are injected to induce porosity.
Oil Shale	Organic-rich black shales (organic content reaches 15–20% by weight exceptionally high) that have not reached the "oil window" and necessary pressure and temperature field due to insufficient burial	Immature. Comprises kerogen: Type II (mixed plant and marine material).	Lothian area of Scotland, Kimmeridge Bay — plays no part in the Wessex petroleum system	Mining. When pyrolysed (heated and retorted) at 500°C the shale will produced oil. Energy intensive. Produces high quality, low sulphur, sweet crude.

(*Continued*)

Table 6.1. *(Continued)*

Type	Geological setting and rock type	Hydrocarbon characteristics and reservoir type	Type area and company	Extraction and processing method
Unconventional Gas				
Shale Gas	Organic-rich shales. The gas is both produced and trapped within the shale	Methane (natural gas) which is trapped in impermeable shale. Can be derived from mixed marine and plant material (Type I and II kerogen)	Barnett shales. North America	Tight gas. Fracking required
Coalbed Methane	Preserved carbonaceous rock formed from plant material	Methane (natural gas) which is trapped in the coal seam.	Worldwide	Through different applications of borehole technology

Figure 6.12. Kimmeridge Bay. View of Kimmeridge clay which has a high organic content and, as is evident from the abundant ammonites, formed in a marine environment.

medium sour synthetic crude oil. According to Suncor's website at http://www.suncor.com this involves the following stages:

- Hot water is used to separate the bitumen from the sands. This step is called extraction and is not required for *in situ* bitumen.
- Bitumen is heated and sent to drums where excess carbon (in the form of petroleum coke) is removed.
- The superheated hydrocarbon vapours from the coke drums are sent to fractionators where vapour condenses into naphtha, kerosene and gas oil.
- The end product is synthetic crude oil. It is shipped by underground pipelines to refineries across North America to be refined into jet fuels, gasoline and other petroleum products.

Figure 6.13. View of Athabasca oil sands mining operations in Alberta. Reproduced with permission of Photographic Services, Shell International Ltd.

6.4 Coal

For a bulk commodity such as coal, the concept of grade used for base and precious metals does not apply. The whole volume of material will be extracted and in some cases shipped directly without pre-processing. Setting up the financial model therefore has to take this into account. IC-CoalEval automates this process and the application includes the following scenario given in the User Guide:

"It is proposed to establish a Longwall section on an underground coal mine to mine out an area of the coal reserves of the mine 2500 m by 5000 m. This portion of the mine has a coal seam 2 m thick on average, so there is a reserve of 25,000,000 m^3 of coal available in the area. The mining layout will be parallel advancing longwalls with 10 m gate protection pillars being left between them. The longwall section will have a called for production capacity of 10,000 tons of mixed quality coal per day. The longwalls will be 231 m wide, gate road to gate road, and the block of coal to be mined is estimated to be 2.5 kilometres long. Taking each gate road as 4.5 m wide the block of coal being cut by each longwall is then 24 m wide by 2500 m long. At a seam width of 2 m with the average density of the coal at 1.35 t/m^3 then there is a resource of 1,620,000 tons of coal to be mined by each longwall with an estimated quality of 90% steam grade, 5% coking grade and 5% anthracite. It is estimated that the dilution will be 5% and the mining recovery will be 90%. The life of each longwall is therefore expected to be some 154 days and moving the equipment to the next longwall is expected to take 16 days, with the longwalls being worked on advance and retreat alternately."

The key form that handles volumetrics and the variables that need to be considered is shown in Figure 6.14.

We have been told the project has an *in situ* resource of 25 million m^3 of coal seam. The *in situ* resource is what is in the ground and will be more than what we are actually able to mine. Based on the geology we can calculate that, given a specific gravity of 1.35, the total tonnage of coal is 33.75 million tonnes. We can now determine how much of the *in situ* ore is mined. To do this we enter the estimated loss of resource due to the selected mining method, in this

IC-CoalEval - Resources

Total Mineralised Vol. (Mcub.m)	25	Coking | Bituminous | Anthracite	
Specific Gravity of Ore	1.35	Practical Yield (%)	5
Losses (%)	10	Wash Recovery (%)	90
Mining Dilution (%)	5	Total Insitu Coking (Mt)	1.69
% Moisture Content	5	Total Coking Mined (Mt)	1.68
		Total Coking Recovered(Mt)	1.51
MTIS Resource (Mt)	33.75	ROM (adc) (Mt)	31.90
MTIS Reserve (Mt)	30.38	ROM (as delivered) (Mt)	33.50
Dilution Mined (Mt)	1.52		

Help OK Cancel

Figure 6.14. Form taken from IC-CoalEval.

case we are using 10%. This is because some pillars of coal are usually left *in situ* for support reasons, often between adjacent longwalls, as is the case in this example. Also the full thickness of the seam may not always be mined for rock mechanics-driven safety reasons, or to reduce surface subsidence problems. The result will be that not all the resource in the ground will be able to be mined and sent to the washing plant.

Next we can enter any dilution factor, in this scenario 5%, which means that every 9.5 tonnes of coal sent to the washing plant will also contain 0.5 tonnes ash (waste rock) i.e. the rock bordering on the coal seam. Having some dilution is normal, as it is often caused by mining conditions, but the key is to keep it to a minimum, as it is often also caused by poor horizon control on the coal-mining equipment.

In coal mining too the moisture content of the coal is important, as it affects both the percentage of the material shipped that is actually coal and the net calorific value of the coal so shipped, as the moisture will need to be boiled off before the calories are available for other use. Here we are using a moisture content of 5%.

This is a multi-product deposit so extra tabs are displayed for each variety of coal, to enter the *Practical Yield* (the percentage of the total coal mined that is of that variety) and *Wash Recovery* (the percentage of that variety of coal sent to the washing plant that reports as product). In this example we are assuming there is 90% bituminous (or steam) coal, 5% coking coal and 5% anthracite. We have been given a plant recovery of 90% for each of these three varieties of coal.

So by clicking on the appropriate product tabs we can enter the percentages of each variety of coal to be produced. For Figure 6.14 we have entered 5% as the *Practical Yield* for coking coal and 90% for the *Wash Recovery* for coking coal. Not seen in Figure 6.14, we have also entered *Practical Yields* of 90% for bituminous coal and 5% for anthracite (not indicated in Fig. 6.14 as it is a static screen shot). The estimated *Wash Recovery* for each rank of coal is set at 90%. When the values for each rank are entered the total *in situ* tonnage, total mined tonnage and total recovered tonnage for that rank of coal are calculated and displayed.

6.5 Coal versus Gas

When building an electricity power station fuelled by coal, objections can be raised because it is perceived to be more "polluting" than one fuelled by natural gas (methane).

Taking the chemical thermodynamic approach you get the following:

$$\underset{\text{Coal}}{\mathrm{C}} + \mathrm{O_2} = \mathrm{CO_2},$$

$$\underset{\text{Methane}}{\mathrm{CH_4}} + 2\mathrm{O_2} = \mathrm{CO_2} + 2\mathrm{H_2O}.$$

Based on enthalpy

$$\Delta \mathrm{H}^{\circ}_{\text{reaction}} = \Sigma\Delta \mathrm{H}^{\circ}_{\text{f(products)}} - \Sigma\Delta \mathrm{H}^{\circ}_{\text{f(reactants)}},$$

then both reactions each produces 44 g/mol CO_2 but coal produces 32.79 kJ/g of heat compared to methane, which produces 50.14 kJ/g. So for the same CO_2 emission, methane produces 1.5 times the amount of specific heat compared to coal. Methane is therefore "cleaner" in as much as more heat is generated per unit of CO_2 produced. It is nothing to do with "pollution".

There is a separate narrative around heat exchange efficiency during the generation of electricity which is greater for methane than coal. Gas-fired power stations run a combined cycle, both a gas turbine and a steam turbine, from the heat rejected from the first process. Therefore, you use much less fuel input to produce the same electrical energy and have proportionately less CO_2 emission and, therefore, less greenhouse gas forming.

Furthermore, coal has inherent moisture so the overall specific heat will be reduced by the enthalpy of vaporisation of water. Shale is usually present so there is further energy loss due to the heat capacity of this material. Of course pyrite is ubiquitous in coal so there is an exothermic reaction there, but SO_2 is a true pollutant and needs to be scrubbed, adding to operating costs.

If you introduce carbon capture and sequestration when considering coal, then you close the CO_2 gap significantly but not in full.

Project Evaluation

Chapter 7

Resource Evaluation

7.1 Introduction

Resource evaluation is a vital component of the project appraisal process. The financial modelling of the project depends on the amount and quality of mineral in the ground and the depth of overburden above it. If these variables are not correctly estimated to a satisfactory degree of certainty, then the subsequent economic evaluation will be inherently flawed. Resource estimation provides the foundation on which the principles of conventional and financial engineering are based. There will always be uncertainty attached to the estimation of the amount and quality of *in situ* valuable minerals, right up until production commences. However, it must be ensured that this uncertainty has been defined as accurately as possible before any investment decision can be made.

Experienced geologists, engineers, and geostatisticians must undertake the technical resource evaluation, as the integrity of the subsequent economic evaluation depends on the reliability of their findings. There are three major techniques for collecting data on the deposit:

- Geophysical techniques.
- A drilling programme, and
- Down-hole exploration.

Drilling provides the greatest resolution and is by far the most important tool in resource/reserve determination. Upon collection, data are processed by standard polygonal techniques and/or more complex geostatistical techniques, in order to construct isopach and isopleth maps. Isopach maps delineate mineralised strata and overburden thickness while isopleth maps illustrate the spatial pattern of other variables, such as resource grades. The following sections summarise the most common data processing techniques used to determine the extent of a mineral resource.

7.2 Sampling

The basic tool during exploration and evaluation is the diamond drill rig (see Figure 7.1). Typical costs per metre would be around $100. The cylinders of core produced are logged and mineralised zones split with a diamond saw and sampled. Preservation in storage of

Figure 7.1. Diamond drill rig, crowns, core and core storage. Preservation of core is essential to support resource estimation, process mineralogy and due diligence. Core diameters are normally produced in a range of sizes designated AQ (27 mm), BQ (36.5 mm), NQ (47.6 mm) and HQ (63.5 mm). Images of core storage and handling reproduced with the permission of Ivanplats (bottom left) and Nkwe Platinum (bottom right).

Figure 7.2. Views of a RC rig at Thabazimbi iron ore mine which is used for grade control aimed at ensuring that product specifications meet requirements (see Chapter 9). Drill-hole spacing is around 10 m, far closer than could ever be expected at the evaluation stage of a project. Reproduced with permission from Anglo American.

mineralised sections of split core is essential for due diligence, particularly post Bre-X (Section 3.4.1).

Once the geological characteristics of a deposit have been identified from diamond drilling, evaluation can be followed up with reversed circulation (RC) drilling. Cost per metre can drop to $20. Features of RC drilling are given in Figure 7.2.

7.3 Process Mineralogy

The Bre-X promoters described the mineralisation as being an epithermal gold deposit (see Figure 7.3) yet independent mineralogical studies queried why was the gold, identified in crushed samples used by the analytical laboratory responsible for the gold assays,

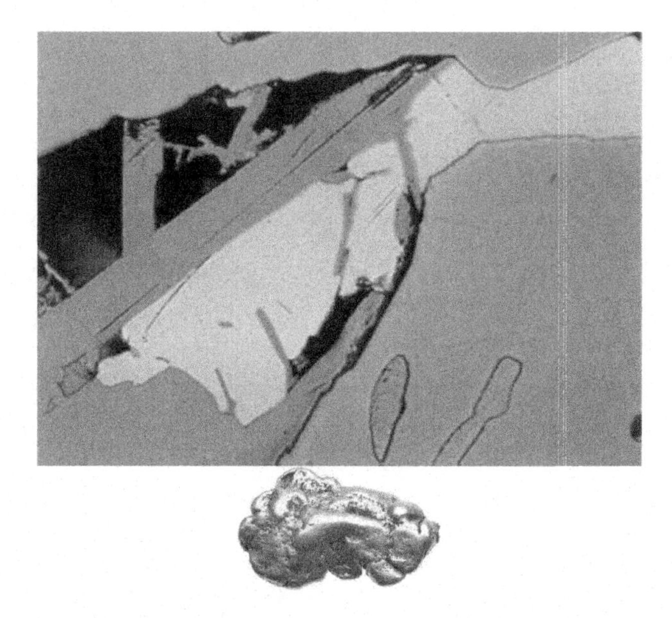

Figure 7.3. Reflected light view of gold inter-grown with pyrite from an epithermal deposit. Black areas are voids which were pathways of mineralising fluids. Horizontal field of view = 0.5 mm. Below, the characteristic appearance of a gold nugget formed in placer alluvial deposits.

in the form of detrital gold with the characteristic morphology of placer gold. The obvious answer was that the samples had been salted with alluvial gold. Primary mineralogical associations were never preserved as all the core was crushed. It follows from this that the diamond drill core should always be split down the centre with a diamond saw and only half the core sent for assay, thereby preserving a continuous representative section of core. Any sub-sampling should be on quartered core and only in exceptional circumstances should intersections through the mineralised zone be sacrificed.

7.4 Concepts Around Geological Continuity

Fundamental to any approach to resource estimation is that there is geological continuity between sample positions. End member

Figure 7.4. View of the Merensky Reef (top left) and the Carlin epithermal gold deposit (bottom right). Continuity of the Merensky Reef is such that sampling is confined to raises that are 120 m apart. The Merensky Reef also has a clear visual marker which is the contact between the pyroxenite (brown) and the anorthosite (grey). Mineralisation at Carlin is associated with the bleached vein systems; it is impossible to predict distribution in the next face following blasting.

scenarios are illustrated in Figure 7.4 with reference to the Merensky Reef and the type area for epithermal deposits in Carlin.

The continuity of the Merensky Reef is remarkable apart from local geological discontinuities caused by potholes and faults. The horizon has been traced for 100s of kms on the surface and drilled down dip to more than 3 km. For Carlin-type epithermal deposits not only is continuity confined to carbon-rich rocks that may be 50 m thick, but also assays for high-grade samples such as that shown in Figure 7.3 from two separate portions of split core are unlikely to provide reproducible results.

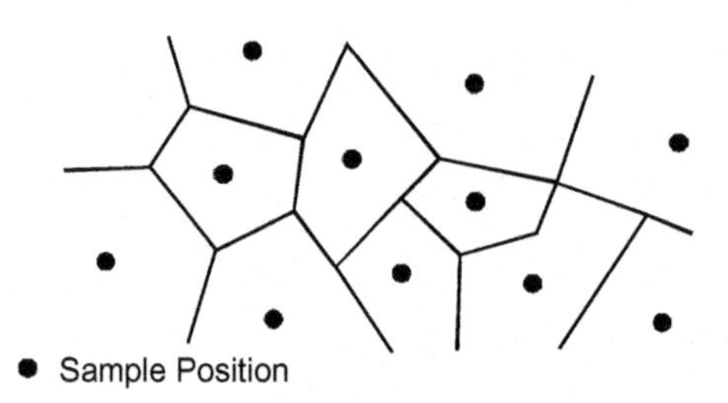

Figure 7.5. Polygons of Influence.

7.5 Polygons of Influence

The historical approach to estimating the grade and tonnage of a mineralised horizon is to divide it into polygons which each contain one drill hole sample (Figure 7.5). This is usually generated by drawing perpendicular bisectors between samples or by expanding circular zones of influence. The variable value of the sample (e.g. grade of valuable element, thickness of mineralised zone, etc.) is then taken as the value of the variable for the entire polygon. The level of uncertainty in the results of this method is generally high, and the reliability of the estimates is predominantly a function of drill hole spacing. Sample variables (metal content in particular) tend to be more variable than block grades, and the results for the whole mineral deposit should be used with caution when planning for selective mining.

The method is useful for volume-weighted global mean grade and tonnages. As sample grades tend to be more variable than block grades it tends to over-estimate the grade of blocks containing high-grade samples and under-estimate those with low-grade samples. If using a cut-off grade it reports too selectively (fewer tonnes at higher grades).

7.6 Inverse Distance

In this method the mineralised horizon is divided into mining resource blocks and the variable is estimated using a weighted combination of proximal samples, with those samples closest to the block having the greatest influence (Figure 7.6). Samples are weighted by

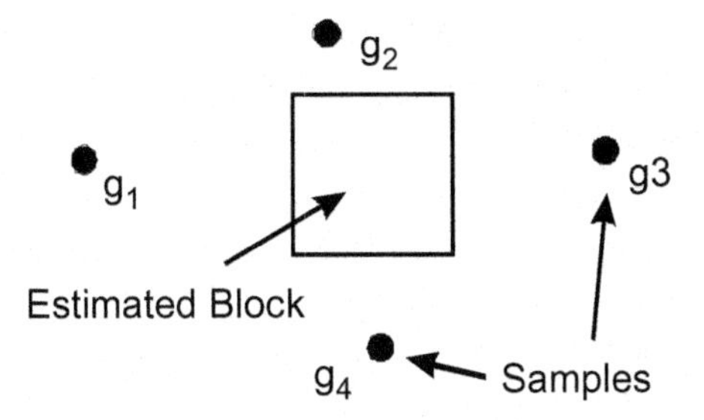

Figure 7.6. Inverse distance method of estimating grade in a block from proximal samples.

estimating their "inverse distance" from the mining block (or inverse distance squared, etc.). The choice of weights introduces a highly subjective element into the estimation procedure and takes no account of the geological structure of the mineral deposit, nor does the method provide any indication of the quality of the estimate (i.e. no quantitative measure of the inherent uncertainty).

Historically complex polynomials were generated to reconcile sampling information available for a mining block which had been extracted and the metal content established. This has the following disadvantages:

- The choice of weights is subjective.
- Weighting does not reflect the geology of the deposit.
- Smooths data without reference to spatial variability.
- Point value set equal to block value.
- The size and shape of the estimated block is not taken into account.
- Does not minimise the estimation variance.
- There is no indication of the quality of the estimation, and
- There is no indication that optimum sample spacing has been used.

The industry was clearly ready to accept a more robust approach to resource estimation.

7.7 Geostatistical Methods

Geostatistics involves a more complex "moving average" technique, intended to provide the best linear, unbiased estimate of the variable in question (e.g. grade). Weighting factors are selected in order to give an unbiased estimate of the true value and to be as close as possible to the true value by minimising the mathematical variance of the estimation.

The field of geostatistics was pioneered by Danie Krige (26 August 1919 to 3 March 2013) a South African mining engineer who worked on the paleao placer gold deposits of the Witwatersrand in South Africa — the technique of kriging is named after him. Krige's empirical work to evaluate mineral resources was formalised mathematically in the 1960s by French engineer Georges Matheron. In 1968 he created the Centre de Géostatistique at the Ecole des Mines de Paris. The application of geostatistics to the challenges of grade control in an active mining operation was promoted in a very practical way by Colin Dixon of the Royal School of Mines in the 1970s. The technique is now widely used and forms a standard part of degree courses in natural resources.

Geostatistical evaluation is a complex task and the assistance of a qualified, experienced consultant should be sought when interpreting geostatistical analyses reported in feasibility studies. The presence of a geostatistical evaluation in a feasibility study, however, is a positive sign and indicates the integrity of the reported resource and reserve estimation. Though not directly reducing the uncertainty of estimates, geostatistics quantitatively measures the level of inherent uncertainty and can also be used to determine the optimum drilling pattern for reserve estimation.

The principal tool in geostatistics is the experimental semi-variogram. This is defined as follows:

$$\gamma(h) = \frac{1}{2N(h)} \sum_{i=1}^{N(h)} [z(x_i + h) - z(x_i)]^2.$$

An interpolated variogram is then fitted to the experimental variogram generated from the primary sampling database, as illustrated in Figure 7.7.

Where sample spacing falls within the range of the variogram, the resource can be classified as indicated (see Section 3.4.1). Close-spaced sampling will delineate a measured resource. When outside

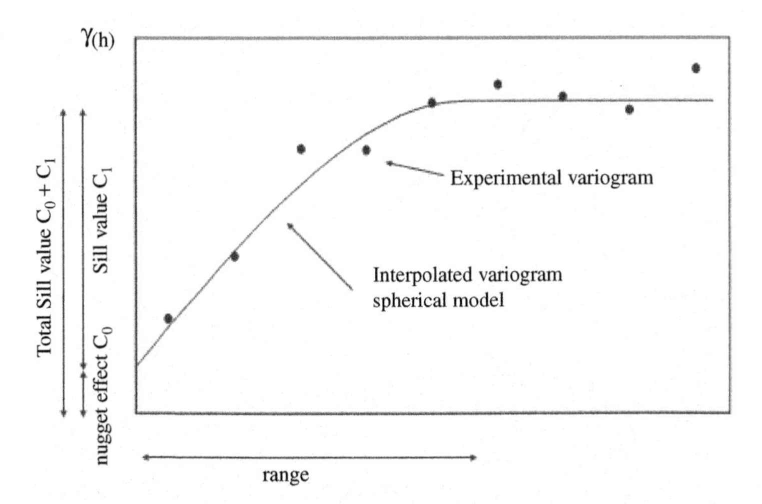

Figure 7.7. Variogram showing the relationship **Y**·(h) which relates the semi-variance of sample differences to distance between samples.

the range of the variogram this is classified as inferred. The allocation of the category is the responsibility of the geostatistician.

Resource blocks are superimposed on the mineralised unit and are then populated with grades derived from proximal samples determined from the application of the associated variogram. The dimensions of the resource blocks are determined by the geostatistician and at the evaluation stage would normally be much larger than the mining blocks established at the production stage.

Where a filter is imposed on a resource block model in which blocks with grades below a set cut-off are excluded from the global estimate, the characteristic set of grade tonnage curves, as indicated in Figure 7.9, are generated. As resource blocks below the determined cut-off are excluded from the resource estimate, the residual blocks will generate a higher overall grade, but there will of course be a corresponding reduction in the tonnage. The boundary condition for the cut-off grade in an underground mine would be the value where mining continuity is lost and the ore body breaks up into blocks that are no longer contiguous. In an open pit operation, where all mineralised material falling within the pit shell has to be excavated,

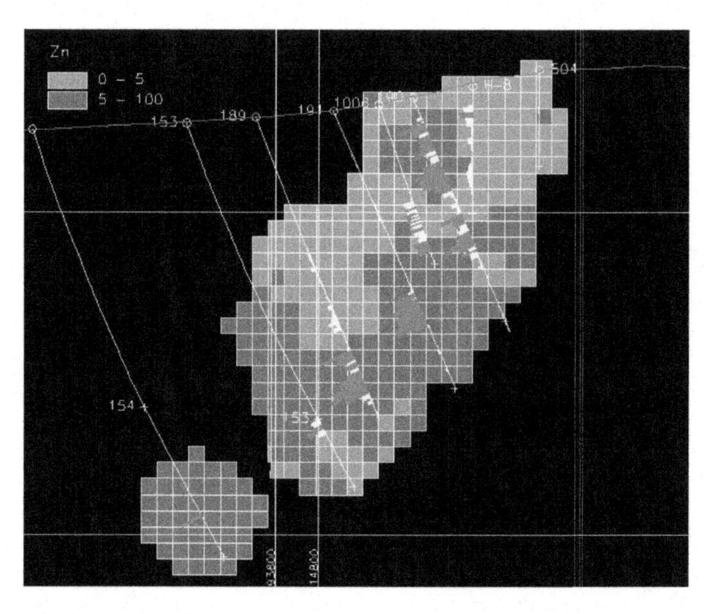

Figure 7.8. Resource block model showing zinc grade estimation. Reproduced with permission of EduMine and taken from Houlding 2015.

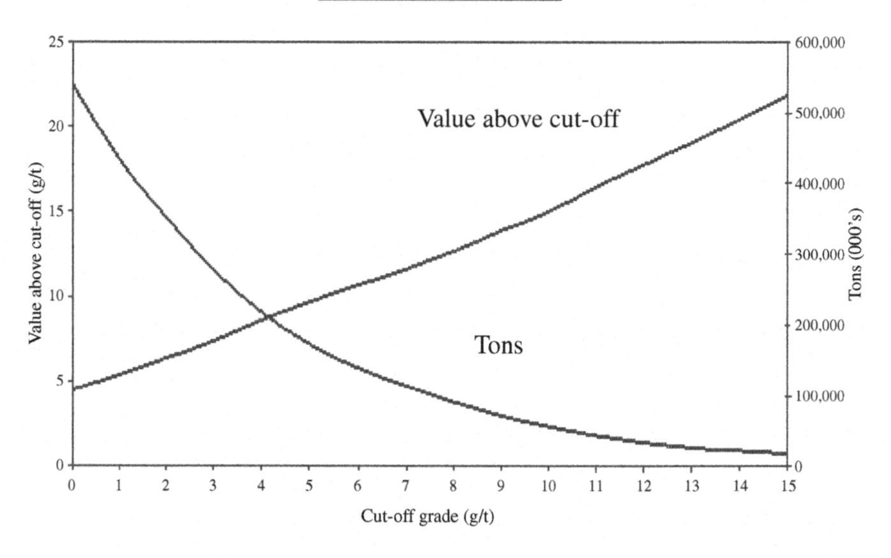

Figure 7.9. Grade-tonnage relationships.

there may be little advantage in attempting to extract preferentially the high-grade zones within the deposit.

7.8 Reserve and Resource Definitions

The codes for reporting mineral resources (see Section 3.4) specify that the financial modelling should normally be confined to the indicated and measured categories. The results would provide a basis for a preliminary economic assessment (PEA) or a PFS (see Section 8.3.1) which would be needed before a resource could be converted into a reserve, and then only if the exercise demonstrated that the deposit is capable of supporting a commercially viable operation. The results of trial mining should permit part of the resource to be classified as measured and therefore potentially a proved reserve.

If there is a component of inferred resource associated with a deposit (by definition, sample spacing is outside the range of the variogram) then further drilling could conceivably convert this to indicated and thereby allow a higher estimate of total metal/mineral content to be incorporated into the financial model. The wider the spacing the fewer the holes needed.

There is therefore a direct correlation between expenditure on drilling and enhancing the valuation of the project. Considerable skill is needed to know when this, in turn, gives diminishing returns — adding to the indicated resource without simultaneously converting that to probable reserve is pointless.

Where evaluation drilling has resulted in the delineation of an indicated resource and where a PFS that includes a pre-funding discounted cash flow model has demonstrated economic viability, then considerable additional value may be added by converting any associated inferred resource into an indicated resource. Very often the inferred resource is much larger than the indicated at an early stage of the evaluation programme so a significant proportion of the potential value rests with the inferred resource. Provided there is transparency in the process a "what if?" scenario can be generated based on plausible assumptions on predicted grade, tonnage, capital and operating costs for a project with an expanded production capacity.

The enhanced NPV that would often arise provides the justification for undertaking the additional evaluation drilling and can be a very convincing supporting narrative in an initial public offering.

7.9 Evaluation Versus Production

There is a fundamental distinction between the point at which you get diminishing returns on drilling at the pre-development stage of a mineral project and grade control in an active mining operation. Theoretically the financial performance indicators between the evaluation stage and re-running the original models with additional sampling should not be different. What is different is that on a day-to-day basis operating staff are making decisions on what benches are ore, waste or low-grade stockpile. The conclusion is that in making an investment decision, the promoters need to understand the production challenges involved in real mines.

We know from geostatistics that once drilling is in the range of the variogram and you have defined the nugget effect from close-spaced drilling in a selected representative area, "in-fill" drilling is not an intelligent use of money. The whole narrative around the Ivanhoe drilling and trial mining site focuses specifically on this issue. The photograph in Figure 7.10 shows the three drilling rigs at 150-metre spacing — agreed by their independent geostatistician.

The purpose of trial mining is to establish variability of grade on the scale of mining as well as to obtain bulk samples for detailed metallurgical test work. The minerals engineers (mining and processing) can only undertake detail design studies on the back of this information. (Incidentally, the task of controlling trial mining must always be done by a mining geologist as detailed documentation of geological context is essential.) The resource reporting codes such as JORC, SAMREC, etc. were specifically set up for evaluation projects and in particular those seeking public funding through an IPO.

The role of the grade control people is to ensure that product specifications meet requirements. In Figure 7.2 of an open pit we see the close-spaced RC drilling, and the challenges of blending through in-pit scheduling is discussed in Chapter 9. No end of confusion is

Figure 7.10. Drill spacing on the Platreef of the Bushveld Complex. The photograph shows three drilling rigs at 150-m spacing — agreed by their independent geostatistician, the qualified person under TSX regulations — which permitted them to "book" the resource as indicated. Reproduced with the permission of Ivanplats Platinum.

caused by companies using the same JORC and SAMREC system of nomenclature in reporting their resources in their annual reports for their active mining operations.

Sampling "support" during mining far exceeds anything that we can expect to achieve at the evaluation stage and here there is a disconnect around evaluation versus operational grade control. Mining companies with active operations are well aware of this, which is why in their annual reports they define their use of "probable" and "proved" reserves in terms of internal guidelines.

There was a sterile debate a decade ago when there was a move to incorporate the resource reporting in the annual financial reports with a value given to a reserve included as an asset on the balance sheet. This seemed to be the same meal ticket that consultants for the petroleum sector have in signing off annual SEC reserve statements (see Section 4.6). It took an accountant to point out that if you used metal price in assigning a classification then the proportion of the mineralisation allocated to a reserve could fluctuate year-on-year, creating wildly different tax obligations.

There is a lack of comprehensive guidance for mining companies under IFRS. The major issues relate to asset recognition and measurement in the treatment of reserves and resources. IFRS permits either historic cost or current value as a measurement basis for assets recognition. Historic cost is currently adopted by most mining companies since IFRS only permits the use of current value if it can be measured reliably. The consequence of this is that the statement of financial position does not give a fair indication of the value of reserves and resources that the mining company has invested in. The alternative approach to asset recognition is based on a fair value measurement but this results in significant volatility in outcome, as it is a function of metal price and operating costs. Historical cost is therefore the more appropriate method for asset measurement.

Chapter 8

Metals and Energy Project Appraisal and Finance

8.1 Introduction

The stages in the development of a mineral project are outlined in Chapter 3 and summarised in Table 8.1 against the corresponding supporting documentation that is needed.

For exploration companies the best approach is to undertake the programme of work with funds being raised through a private placement. This is not uncommon for early stage exploration and, of course, there would be no liquidity as changes in shareholding would be directly between two parties rather than exchange-based trading. As the exploration success allows the project to evolve, the original investors will want to take the company public. This process of bringing a project into production is inherently complex and promoters and investors get hopelessly confused about the interrelationships between the funding options and the stages in the technical evolution of their projects. Advisors add a veneer of mystique over the process to justify large fees. This chapter aims to provide a clear road map for the development of a mineral project, although the process is not that different for a petroleum project. AIM's guidelines incorporate mining, oil and gas companies in the same document.

8.2 Role of Stock Exchanges

There is a link between "liquidity" restrictions in selling shares in a natural resource company and what are classified as "Junior" or "Small Cap" and "Major". The argument is that once "Junior" shares

Table 8.1. Nomenclature of documentation generated in mineral project development.

Stage	Technical Studies & Contracts	Stock Exchange Requirements		Investment Bank			Permits
	Format given by Professional Organisations	Toronto Stock Exchange	London Stock Exchange	Debt	Equity	Trading	
Pre-IPO	Geological potential. Generated by Competent or Qualified Person						Prospecting
Listing	JORC, SAMREC, etc. Codes, Preliminary Feasibility Study. Generated by Competent or Qualified Person	NI43-101 format technical report/ prospectus	AIM Note for Mining, Oil and Gas Companies Prospec-tus. Role of NOMAD		Underwriting Analyst reports		Exploration
Joint Venture Agreement	Due diligence based on data room. Geological potential. Generated by competent or qualified person				Preliminary Information Memoran-dum		Exploration
Funding	Full technical feasibility study. Front end engineering design. Environmental impact statement			Information memoran-dum, term sheet.	Rights issue	Hedging strategy. Options	Mining
Construction start-up	Procurement, construction and management. Issued for tender/ construction. Handover certificate			Construction monitoring, economic completion test. Production monitoring			Mining

are acquired by a fund manager they would have difficulty disposing of a shareholding in these companies without there being a significant negative impact on share price (see Section 3.8). A more objective basis for considering liquidity would be the choice of stock exchange. Professionals understand that natural resource projects should be seen as part of the cycle from exploration though evaluation into pre-development, development and then production. Disposal of shares at the inflection points shown in Figure 3.5 would be a perfectly rational decision to make, as there may well be a corresponding demand at that point.

The TSX is recognised as one of the leading natural resource markets and companies listed cover the whole spectrum from exploration through to production. It has a main listing for natural resource companies which, in general, share as a common feature projects that are in development and production.

The TSX Venture Exchange specialises in natural resource companies at the exploration stage and evaluation stage. Choice of listing is a decision the Board of Directors would take. Migration from the Venture Exchange to the main TSX would imply that projects are progressing towards the development stage. Failure to meet such market expectations would not help share price. Liquidity in the shares of a company is not related to size. An exploration company that has positive news flow would expect to see a rise in its share price and there would be a demand for these shares.

The share trading systems used by the TSX make no distinction between different categories of natural resource shares. The trading platform for the main market is exactly the same as that for the TSX Venture Exchange. Lack of liquidity will arise if market sentiment is depressed and sellers exceed buyers with the result that direct intervention by brokers may therefore take the place of online trading systems. Matching sellers with buyers in this type of market will obviously lower pricing and there may be a time delay before a sale is secured.

The main constraints on selling shares in an exploration company listed on the TSX Venture Exchange might occur where acquisition of the shares involved specific caveats against selling, such as when a director of the company has acquired a significant shareholding. The director would be expected to demonstrate confidence in the future of the company and maintain the holding regardless of the vicissitudes of the market.

8.3 Preliminary Feasibility Study

8.3.1 *Elements and characteristics*

A successful IPO will be expedited if the project has advanced to the point where an indicated resource has been delineated. This permits a PEA or a PFS to be undertaken. This usually takes around six months to complete and should cost in the range of $250K to $500K. It cannot be used for major programmes of capital expenditure and is not acceptable to banks for debt finance. For good projects, capital and operating cost estimates on an initial conceptual project design need not be more accurate than ±20%.

In addition to the resource estimation, the technical components of a PFS should also include the results of process mineralogical studies and metallurgical test work. Post Bre-X (see Chapters 3 and 7) the process mineralogical study is essential. A pre-funding DCF model should form part of a PFS which would include tax and royalty provisions. It would always be an advantage if as part of the exploration phase an environmental baseline study had been completed (Chapter 4).

A PFS should normally be undertaken by an interdisciplinary team where good cross-communication is present. A range of different scenarios can be considered (as outlined below) and provides the basis for directors to conclude whether or not they should proceed with a project. A PFS also provides the baseline against which future investment decisions are made. It helps identify where further expenditure would add most value and often this shows up the need to add to the resource base.

8.3.2 *Cost estimation*

Estimation of costs is required at the PFS stage after ore reserves are determined, but before major capital is committed. At this stage

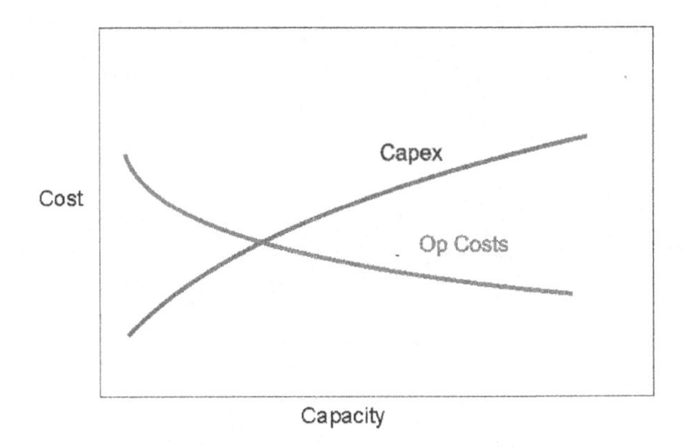

Figure 8.1. Relationship between cost and capacity.

there is insufficient technical knowledge available for accurate estimation of costs. Estimates are based on calculated average costs of existing mine projects and ±20% accuracy will suffice.

There are two types of cost: capital costs (Capex) which are costs in a particular year that will produce benefits in later years, and operating costs (OpCosts) which are costs which only produce benefit for that year. The relationship between cost and capacity is given in Figure 8.1. Capex increases with increased capacity but OpCosts generally decrease with increased capacity due to the economies of scale.

Even at the PFS, estimating costs do offer challenges. A starting point might be to use an analogue but there are online mining cost estimating systems such as CostMine available at:

http://costs.infomine.com/.

Once a scenario is costed, then flexing of costs can be achieved by using the following empirical formula:

$$\left(\frac{\text{Cost 1}}{\text{Cost 2}}\right) = \left(\frac{\text{Capacity 2}}{\text{Capacity 1}}\right)^{0.6}.$$

For a base case mining operation with a production of 0.94 Mt/annum which is ramped-up to the target that the mine feels they can achieve of 2.5 Mt, this then permits a commensurate reduction in OpCosts. If in the base case it is given as \$44/tonne run-of-mine, then the formula predicts a revised cost of \$24.58/tonne run-of-mine.

An example of the approach needed for more detailed cost estimates is given in Section 9.4.2.

8.3.3 *Asset optimisation*

A PFS and a FTFS play key roles in defining the transition from an indicated resource to a probable reserve, and form an integral part of the process of applying the codes for reporting mineral resources as discussed in Chapter 3. The most important difference between a PFS and a FTFS is that as there is more flexibility associated in constraining technical variables in a pre-feasibility study, it is much easier to undertake scenario analysis. This permits optimisation of an asset and therefore the NPV using a simple financial model based on a fixed discount rate. The size of a mining project can then be constrained by the time value of money, the choice of cut-off grade on reserve block models, and the impact of operating and Capex arising from economies of scale. This is demonstrated in Figure 8.2.

The rate of production for very long life assets may be scaled up, thereby reducing OpCosts and allowing lower grade ore to be exploited. Conversely there will be a boundary condition when increasing the cut-off grade as mining continuity is lost.

Once the deterministic "sweet spot" has been selected based on a fixed metal price, then a MC simulation as outlined in Section 2.4 based on grade, plant recovery, operating and Capex would generate a probabilistic distribution of NPV that provides a more realistic perspective of the robustness of the project. Changing metal price may allow the project to be re-optimised as, for example, open-pit design and the capital to operating cost mix may be revisited. Capital has to be spent on open-pit push back to provide access to deeper ore or the investment made in going from open pit to an underground operation.

Volatility in prices can cause difficulty in establishing a stable operating cost and capital cost mix at the PFS stage. The technique of ROV discussed in Sections 4.8 and 10.3 demonstrates how price volatility can be formally modelled using the principles of quantitative finance. Simply flexing price is a pointless exercise as it generates self-fulfilling outcomes.

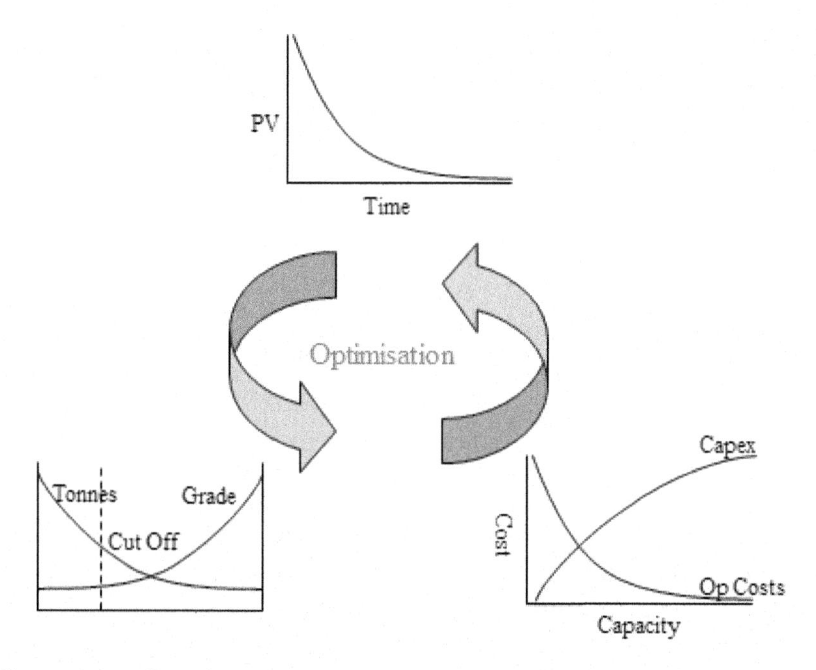

Figure 8.2. Size of a mining project is constrained by the time value of money, the choice of cut-off grade on reserve block models, and the impact of operating and Capex arising from economies of scale.

Cut-off grade is far more likely to be determined by geological conditions and mining methods than what happens on the London Metal Exchange (LME) or currency markets. Reducing cut-off grade in a copper mine with the ore body coincident with the envelope of mineralisation (which is the case for most big porphyry deposits) does not "create" more copper.

8.4 Joint Venture

It can be a challenge for the Junior to secure the interests of a Major to enter into a JV and the investment banks can play a role by generating a Preliminary Information Memorandum. This includes due diligence on the legal and financial standing of a company. It incorporates description of licences and outlines the structure of local JVs. It also incorporates a technical report prepared by the Junior or independent consultant.

The structure of a JV is given in Section 3.8.3. In addition to the legal agreement that will be entered into there has to be a process of due diligence: the guidelines are given in Section 4.3.

8.5 Feasibility Studies

8.5.1 *Introduction*

If a mineral deposit is classified as a reserve, then independent technical and financial studies must have been undertaken and have demonstrated that the deposit can support a commercially viable operation. Such a "Feasibility Study" under paragraph 40 of the JORC code is defined as:

"a comprehensive technical and economic study of the selected development option for a mineral project that includes appropriately detailed assessments of applicable Modifying Factors together with any other relevant operational factors and detailed financial analysis that are necessary to demonstrate at the time of reporting that extraction is reasonably justified (economically mineable). The results of the study may reasonably serve as the basis for a final decision by a proponent or financial institution to proceed with, or finance, the development of the project."

Feasible simply means possible, but a feasibility study (according to the Collins Concise Dictionary) is a study designed to determine the practicability of a system or plan. The technical guidelines for the term are therefore consistent with common usage.

For an engineer who has spent time with a nationalised industry the connotation behind the term feasibility is to determine if mining is technically possible and that revenue generated covers OpCosts. The justification for the capital investment in the equipment needed for the task would not necessarily have been part of the remit. Honest differences in perspective then occur much to the frustration of operating staff who feel that mine development is being restricted by the business and finance divisions. Even when mine planning does take into account the need for return on investment in equipment, preproduction, underground development or pre-stripping in an open pit, the choice of discount rate in a financial model is often jealously preserved by the business and finance divisions who set the corporate cost of capital.

To add to the semantic confusion, a company commissioning a "Feasibility Study" from a group of consulting engineers at a cost of up to a million dollars will not be best pleased if the conclusion is that the project is not feasible. Furthermore there are a range of different types of engineering studies used in the industry and these are outlined below. This approach to conventional engineering needs to be seen within the context of the funding options available to the promoters that includes a combination of debt and equity.

Any investment in a mining system needs to generate a revenue stream that not only covers the cost of extraction, but also a return that exceeds the discount rate selected. Based on these criteria (where the quantitative approach generates an IRR less than the discount rate) then it must be concluded that it would not be feasible to proceed with the investment as mining is not a commercially viable option.

8.5.2 *Elements and characteristics*

A FTFS would typically take around 12 months to complete at a cost of around $1 million. The co-ordinating director would want to mobilise the advisory investment bankers who, in turn, would want to alert their independent engineers. Estimates of the grade and tonnage used in a FTFS may well be derived from the same database used in a PFS. Detailed conventional engineering cannot create metal that was not formed from natural geological processes. Given that once sampling from surface drilling is within the range of the variogram it may well be concluded that additional geological information on the scale of mining is needed. This provides justification for undertaking a programme of trial mining, which would normally be under the control of a mining geologist, as sampling should be undertaken in conjunction with meticulous mapping. This provides constraints on the approach to bulk sampling that can then be used for pilot plant treatment.

8.5.3 *Engineering design stages*

Given that a FTFS would seldom conclude the project is not feasible, it should be considered as essentially a framework for the engineering at the "procurement, construction and management" contract (EPC&M) phase or "lump sum turnkey" contract. While the FTFS is likely to be associated with a pre-funding DCF model, this offers no guarantee that a project is commercially viable. The capital cost estimates will, however, provide the basis for determining the amount of capital needed for construction (both debt and equity).

For large projects seeking debt (also referred to as project finance) the loan will be repaid from cash flows generated by the project. Where there is no recourse or only limited recourse then there are no tangible assets until the operation is in production. Banks are exposed to risk and will take a conservative view. Major mining companies have production experience not readily available to a Junior and a proven track record is a big advantage in securing both debt and equity finance.

The document the bank will generate is the Information Memorandum which outlines the terms of the loan. Protection will be sought against falls in metal prices and a hedging programme outlined. The presence of a hedging programme would of course limit the benefit for the Company in the event of a significant increase in metal price. So a balance must be struck. The amount of metal hedged tends to be limited to no more than 40% of total production. The FTFS will be incorporated into the Information Memorandum as an Appendix which is never described by investment banks as a "Bankable Study" and when the term is used by a Junior it does not inspire confidence.

Project finance allows further optimisation of NPV through gearing and the influence of WACC (see Chapter 3).

Actual construction requires the completion of front end engineering design (FEED). While this process could extend from 24 to 36 months and cost in excess of $10 million it may well overlap with the start of construction. The construction phase involves the following contractual certificates:

- Issued for tender (IFT).
- Issued for construction (IFC) drawings/deviation to be authorised by client, and
- Handover certificate.

Even for a well-managed construction project, final Capex are unlikely to be better than within ±10% of the original estimates. This intrinsic uncertainty needs to be reconciled with provision for contingency in determining funding needs. Contingency should be used to cover unexpected circumstances and not the inherent uncertainty of the original capital cost estimates.

Construction monitoring will be undertaken by independent engineers who will report progress to the lenders based on management systems such as MS Project© used by the engineers implementing the EPC&M contract. These systems will be used to report actual costs to the lender. Banks must be aware of any cost overruns and ensure that the money provided by the lenders is spent according to the development plan. The bank's independent engineers therefore have the important responsibility of highlighting the potential for any overrun in time and cost.

After construction has been completed and the project commissioned, the bank's independent engineers will undertake the mechanical completion or economic completion test. This confirms that the project has been built as specified in the loan agreement and is capable of performing according to the development plan.

Once the project is in operation the production monitoring will be undertaken by the bank's independent engineers to verify the ramp-up in production and OpCosts, and report progress to the lenders. If performance and OpCosts reconcile with those predicted in the original FTFS, then the corporate guarantees to the bank providing project finance may fall away and the lending goes non-recourse. Anecdotal evidence suggests that this reconciliation is achieved in around only half of new mining projects.

8.6 Project Finance Parameters

Input information is required to set up the financing structure of the project including the amount of debt and equity, interest rate and repayment schedule. This will include the following:

- Capital structure. The debt/equity ratio and the size of debt will be decided by the lender. It can be expressed as a percentage of the total financing requirements that will be funded as debt. The optimum draw-down period for the debt funding will be agreed between the project sponsor and the lender, and may be drawn out over as long as the first five years of the project.
- Loan type and repayment schedule. The schedule for loan repayment needs to be established in order to complete the cash flow model. The number and size of loan repayments will be negotiated between the lender and sponsor, as will the grace period, if any, before repayments must commence. Loan repayments can be made in equal instalments (straight loan) or made proportional to the production rate (production loan). There will be other cash flows associated with organising the project finance that must also be included in the early years of the model. These include an up-front fee by the bank for arranging the loan (a percentage of the total loan available), a commitment fee (an annual fee charged on the amount of the loan that has not been used), fixed charges (for agents' fees, legal documentation, independent reports, etc.) and contingency to act as a cushion against unexpected cost rises, etc. (a percentage of the total required funding).
- Loan interest rate. This is the annual rate of interest on the debt as set by the lender.
- Return on equity. This is the annual expected return on equity invested as funds. It can be calculated by a variety of methods including the CAPM and is often linked to the overall company gearing of the project sponsor.

IC-MinEval has a "Funding" tool which uses Goal Seek (an Excel© tool) that searches for a value which satisfies the conditions defined in the above scenario.

There are three types of financial ratios used in project finance, all of which analyse the project cash flow. These can be split into:

Annual cover ratios

- Interest cover ratio (available cash flow/interest rate).
- Principal cover ratio (available cash flow/principal) or (available cash flow — interest payment)/principal payment, or
- Debt service cover ratio (available cash flow — debt service)/debt service.

Present value (PV) cover ratios

- Loan life ratio (LLR) NPV of cash flows after funding during loan period, discounted by the interest rate, or
- Project life ratio (PLR) NPV of cash flows after funding during entire project, discounted by the interest rate.

The difference between LLR and PLR is a measure of the residual cash flow cover.

Overall ratios

- Debt: equity ratio projects with high market risk, typically 50:50 or 60:40, those with lower risk 75:25, or
- Residual cover PV of post-loan maturity cash flows/maximum loan.

As the only tangible asset is the unmined ore, the repayment schedule would be structured such that after the loan has been repaid there would be residual mineralisation, normally at least 30% of the original resource. This is known as the reserve tail.

The application of these performance indicators is covered in Section 10.5.

8.7 Case Histories

8.7.1 *Coal washing plant optimisation*

The major advantage of setting up a financial model of a coal project is that an alternative to where potentially all of the material mined is the product and so the run-of-mine "grade" would be 100% can be considered. Downstream processing can be designed to produce different product steams and a financial model is needed to optimise coal yields.

For example, an open-pit coal mine with a capacity of 200,000 tonnes mixed quality coal per annum might have a resource that has an estimated quality of 70% steam grade and 30% coking grade (also known as practical yields, see Section 6.4). If the first stage pre-treatment processing plant wash recovery is 70%, the cost of mining coal is \$8.00/tonne, the cost of mining waste is \$2.00/tonne and the stripping ratio of coal:waste is 1:6, and the total cost of mining one tonne of coal would be \$20.00 per month.

If the OpCosts for the first stage pre-treatment processing plant wash are \$5.00/tonne, the additional costs of producing separate streams of steam and coking grade coal are \$2.00/tonne and \$5.00/tonne respectively in the downstream processing. Assuming 100% recovery, the overall processing costs per month are \$210,000 and \$196,000 respectively.

Based on price per tonne of steam coal of \$50.00 and that of coking coal of \$120.00 then the overall operating margin is \$4,534,000 per month. If tests showed that when the first stage pre-treatment processing plant was operated such that only steam coal was produced then wash recovery would increase to 90%, and the potential operating margin drops to \$3,640,000 per month. Running the plant to produce premium coking coal therefore improves the potential operating margin, notwithstanding the lower pre-treatment wash recovery, compared to producing only steam coal.

As the price of the different product streams changes so will the need to adjust the product mix. Given that the market for coal is extremely complex, multiple scenarios may need to be modelled and the processing plant designed to allow the necessary flexibility. A typical coal mine may have a dozen different product streams and balancing recovery with price to optimise performance in a coal washing plant is not a trivial task and impossible without a financial model.

One of the fundamental constraints on establishing a viable business in any mining operation that produces a low-value bulk commodity such as coal or iron ore is that the only way you can generate a material investment return for a holding company is through volume. AngloCoal's Zibulo Colliery room-and-pillar underground operation and their Landau open pit are both producing plus 3 million tonnes of coal per annum.

Operators get confused between an operating margin based purely on cash flow and return on investment. The moment any attempt is made to justify a significant capital investment in mining equipment the investment in terms of NPV cannot be supported. To make an investment in a processing plant would require reserves that would sustain at least 15 years of operation.

8.7.2 *Iron ore*

Setting up a financial model of an iron ore project conceptually involves the same approach as that used in a coal project (see Section 6.5). As can be seen in Figure 8.3, dilution is inevitable during mining but there will be fixed mining costs. The price of iron ore is quoted as "US cents per dry metric tonne unit (dmtu) x iron content". At a price of 80 cents "dmtu" and for a product containing 64% (the iron content of pure haematite is 69.94%, see Section 5.9.2), this would give a price per tonne of ore of $51.20/tonne.

Offsetting that are very stringent specifications established by the steel producers with penalties for SiO_2, Al_2O_3 and P. The maximum P content can be as low as 0.04%. As a consequence, iron ore operations need to implement a sophisticated grade control system starting from block modelling, in-pit blending and optimisation of plant recovery.

This would generate the revenue profile indicated in Figure 8.4. In this scenario there is a penalty for ore with an iron content that drops below 62.5% and ore below 60% is not accepted. To optimise performance in the plant the recovery curves (Figure 8.4) also need to be considered. The distribution of NPV based on a nominal capital expenditure for a push-back in an open pit is given in Figure 8.5. This also takes into account not only the cost of capital, stripping ratios and mining dilution as well as the impact of penalties, notably phosphorus, but also the 62.5% boundary condition and the 60%

Figure 8.3. View of a working face at the Sishen mine in South Africa showing the outline of potential mining blocks. The dark brown horizon is high-grade haematite and the grey is quartz-rich BIF. The block on the left will include 50% dilution, that on the right 5%.

limit. There is also a separate issue around differential pricing of fines versus lumpy.

Optimum NPV is derived from a recovery of 85% which predicts a grade of 62.5% iron in the product. This is just on the penalty limit which is probably not coincidental.

The common quotation of iron ore per tonne is normally with reference to 64% iron. There is an analogue with the oil price which is based on Western Texan Intermediate (see Section 4.7). Only a small proportion of oil trade includes this price benchmark. The reference price of 64% iron may not reflect the bulk of iron ore traded. As iron ore prices fluctuate, producers would adjust their operating conditions but these would need to be based on yield optimisation curves. The 64% benchmark may well be adjusted upwards under pressure from major consumers such as Chinese smelters concerned about environmental constraints. This will then impact on iron ore production cost models.

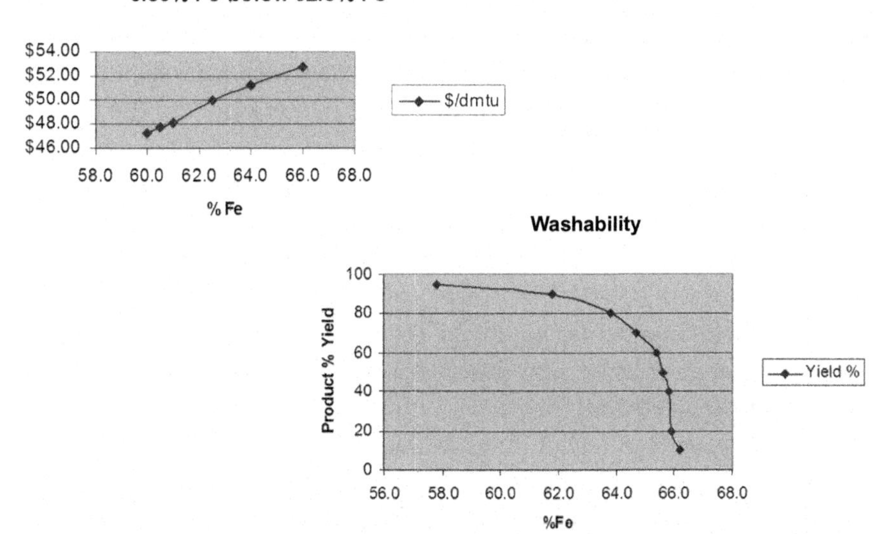

Figure 8.4. Iron ore revenue assumptions −6 to +1 mm Sishen ore recovery curves.

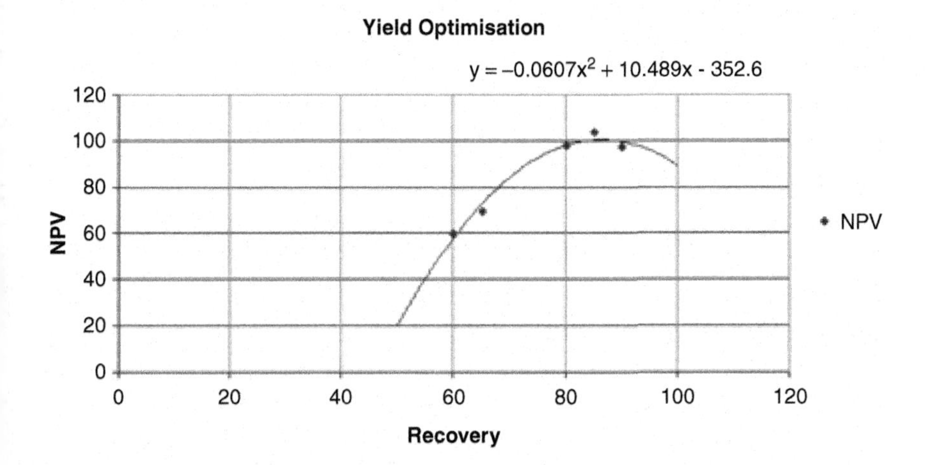

Figure 8.5. Yield optimisation curve derived from Figure 8.4.

As is the case for coal the commercial viability of developing an iron ore project depends on the availability of infrastructure. The dedicated Sishen rail system comprises 861 km of line. Each consignment comprises 216×100 tonne wagons/train with a travelling time in one direction of 18 hours. The port at Saldanha can handle 30 Mt per annum. Alternatives have been considered such as slurry pumping, particularly when magnetite-bearing ore has to be crushed and ground as part of a liberation process or where it is a by-product.

The primary ore at LKAB's Kiruna iron ore operation in Sweden is magnetite, as is the by-product of copper mining at Palabora in South Africa. Magnetite based on stoichiometry has a higher iron content than haematite and in dmtu terms is therefore worth more per tonne than haematite.

The magnetite associated with layered mafic intrusions is normally intergrown with ilmenite and has ulvospinel exsolved on the octahedral plane so is useless for steel production. It is, however, a source of vanadium (see Section 5.9.2).

LKAB produces magnetite from skarn and undertakes an oxidation stage on the fines and sintering that locks in a thermodynamic, and therefore pricing, advantage. Apatite has to be removed using flotation and the same occurs at Palabora which exports magnetite fines to China. In 2013 magnetite was as important a source of revenue at Palabora as copper. Industrial Development Corporation are investing in a pyrometallurgical process to produce metallic iron onsite to enhance value.

8.7.3 *Transition from open pit to underground*

The determination of the transition from surface to underground mining requires careful timing and integration of technical planning and development with a matching investment strategy. The process of undertaking PFS and full engineering planning studies should be initiated in a timely manner, otherwise the sponsor will incur higher costs in attempting to speed up development once the investment decision has been taken. The consequence will be a serious dip in production on changeover. This is illustrated in Fig. 8.6.

Given the challenges in implementing block caving there is a high risk of overestimating how much can be recovered from underground

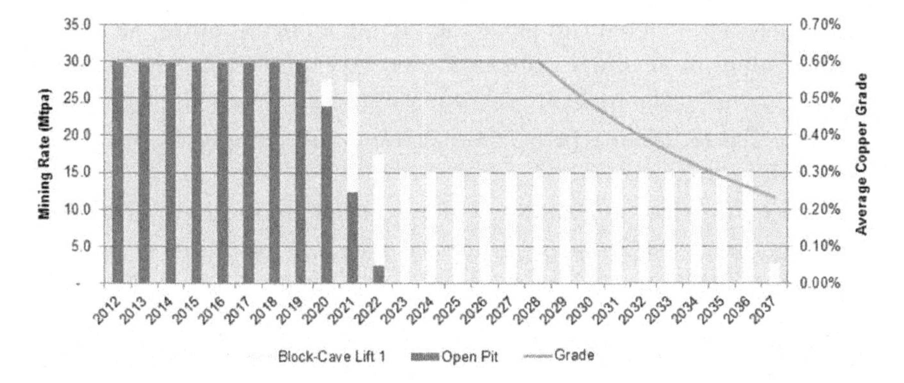

Figure 8.6. Production profile in the transition from surface to underground mining. As the pen-pit matures, the number of working faces becomes limited so production drops and this needs to be compensated for in the ramp-up of block caving. This is conceptually simple but technically challenging. It requires a large design effort, high quality construction and rigorously managed operations.

based on original ramp-up profile assumptions as caving is initiated. This has profound implications when planning matching investment needs and it is the interface between conventional technical engineering and financial engineering that forms part of this study. An integral part of this study was the development of models that consider the impact of unexpected delays in meeting production targets on the original financial performance indicators, such as NPV. These can be ameliorated through the use of innovative funding options aimed at reducing the cost of capital. The impact on reserve tail caused by dilution at underground drawpoints in block caving operations when the original open-pit side wall suffers failure needs to be built into the models, as it will influence the choice of repayment period (see Section 10.5).

8.8 Discussion

For an IPO the prospectus must have a technical report prepared by an independent consultant (Competent Person's Report), which in turn may include all the elements of a pre-feasibility study. It may not be necessary to have a full technical feasibility study completed if funds are to be used for, say, additional drilling or trial mining.

A PFS can provide the basis for raising equity funding for additional evaluation drilling. It cannot, however, be used to raise enough funding for a major programme of capital expenditure and it is not, owing to the remaining project uncertainty, acceptable to banks for raising debt finance.

Project finance investment will be based upon an Information Memorandum, which incorporates a FTFS. Optimisation undertaken at the PFS stage is treated quite separately from the financial engineering that follows using project finance, which is essentially designed to optimise the capital structure. As the gearing changes so does the NPV. The obvious reason why the NPV increases with increased levels of debt is because of tax relief on interest and because the cash flows are being discounted, reflecting the impact of cheaper debt finance.

Putting together a JV requires a valuation of the assets of the Junior as a basis for the determination of the vend-in conditions for the Major and a clear set of targets to identify when trigger points are reached. Usually this is linked to the delivery by the major of a PFS and a FTFS. These need to be defined with rigour if disputes are to be avoided. In the event that the major decides not to proceed, a claw-back clause is often included in the terms of the original JV agreement.

Estimates of the grade and tonnage in the FTFS may be no different from those used for pre-feasibility but may well include the results of trial mining, bulk sampling and pilot treatment. This level of technical information would not usually fall within the scope of a pre-feasibility study. Permitting and EIA also need to be in place (see Section 4.1.1). A FTFS will also provide the verifiable cost and performance predictions. These are needed for the economic completion tests when operations funded from debt go from recourse to non-recourse.

In providing debt a bank would also want to lock in revenue, so that the proportion of income needed to meet interest and debt repayment is secure. Of course the amount of debt that a project can carry is dependent on performance indicators (debt service coverage and ratios such as loan life, project life, reserve tail, interest cover, principal cover and residual cover). The risk for the shareholders is that hedging imposed by the banks could lock them out of significant revenue from the project, because they receive less benefit from any major increase in metal prices.

This professional interaction is tracked through to the role of engineering, procurement and construction contracts within the context of financial engineering and optimisation of debt and equity at the funding stage. This in turn links back to the impact that delay has on financial viability at the construction and development phase, and reconciliation between initial design expectation and actual performance during production ramp-up. Finally, the role of the economic completion test becomes material before lending goes non-recourse.

Promoters with purely technical backgrounds however, often understand poorly the distinction between technical appraisal and financial engineering. Linking the sophisticated hedging strategies that are required when securing project finance to the technical appraisal and exploitation of the underlying assets, is highly complex. Correct structuring of the finance and hedging requirements does significantly enhance the value of mineral projects. For example, gaining access to funding may be value-additive in itself as it provides greater clarity over project timetables. A simple single year delay in moving forward at feasibility stage could erode 5–10% of value purely on a time value of money basis. Raising the finance itself may also be value-creating through the signal to the market that the project risk has been reduced significantly.

Chapter 9

Minerals Engineering

9.1 Introduction

A mine open-pit blast is just the visible manifestation of what is a complex process from *in situ* metal such as gold to a final Doré bullion bar.

It is quite an art for mining engineers to get blast-hole spacing and loading of explosives just right. You want to get optimum fragmentation of ore so most of that material goes through the grizzly at the primary crusher, while ensuring you do not waste explosives by breaking waste rock too small nor having to do secondary blasting on the waste rock. At the same time you do not want to physically displace the rock mass significantly as grade control is based on ore remaining in its original spatial location.

The geologists have to delineate mineralisation and determine zones above cut-off grade that demarcate ore to be sent to the plant. There may be a low-grade component which may be suitable for heap leaching. The mining engineers have to ensure that after blasting there is no misclassification which in turn links to truck scheduling. The plant superintendent has to discharge minimum quantities of gold to the tailings constrained only by recovery curves. Plant recovery is among other things a function of head grade. Expertise is important at all levels from the operators upwards. In the plant the complexity of the system's behaviour and associated relative lack of automation makes operator experience essential for achieving good recoveries and grades.

Figure 9.1. Open-pit blast at the Nkomati mine in South Africa.

9.2 Metal Reconciliation

In a modern mine there must be a metal reconciliation system, and some really robust inter-disciplinary exchanges can take place at the monthly planning meeting. The mineral process engineers attribute lower than expected ounces recovered to low grade in the plant feed. The miners say the geologists gave them the wrong information and that *in situ* grades were less than predicted, and in any event the mineral process engineers have let plant recoveries drop. The geologists say that excessive dilution took place during extraction which is why there was low head grade and ore was sent to the waste pile (and waste sent to the primary crusher). The geostatistician just smiles at the failure of his/her colleagues to understand the inherent uncertainty associated with initial drill-hole sampling.

A starting point in the process of metal reconciliation is often in the treatment of mining dilution. It is seldom if ever possible to mine only the ore. When recovering the ore there will always be some waste rock (see Figure 9.2). The amount of dilution is usually expressed as

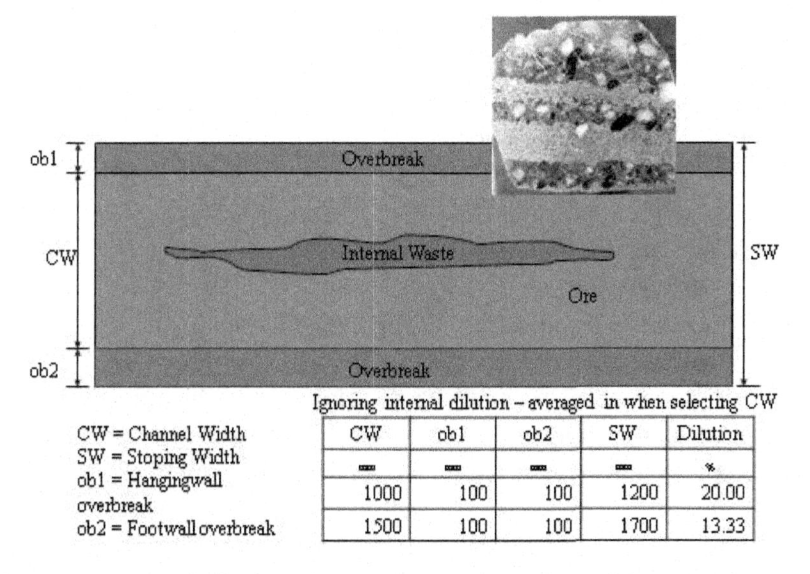

CW	ob1	ob2	SW	Dilution
[illegible]	[illegible]	[illegible]	[illegible]	%
1000	100	100	1200	20.00
1500	100	100	1700	13.33

Figure 9.2. Calculation of dilution assuming the density of the ore and waste is the same. The inset picture is a section of Basal Reef from Free State Geduld Gold Mine in the Welkom Goldfield. Pebbles are around 3 cm in size and are associated with the gold mineralisation, while the brown sandstone is internal waste.

Table 9.1. Mass balance in a gold-mining operation.

Variables	Tonnes	Grade (g/tonne)	Content (grams)
Ore	100	10	1,000
Dilution at 10%	10	0	0
Total to plant	110	8.33	1,000

a percentage. External dilution is derived from outside the ore zone and internal dilution derives from waste bands in the ore zone. The volumetric impact is shown in Figure 9.2.

The object of the exercise is to send ore to the processing plant but, when mining, it is impossible to separate perfectly the waste from the ore. Inevitably some waste gets mixed up with the ore and is sent to the plant. This is "dilution", usually expressed as a percentage and calculated as given in Table 9.1. Dilution material is usually assumed to have no valuable content (which is not always the case) and has the same density as the ore.

The impact of dilution needs to be accommodated in a financial model. For a scenario with a fixed capacity, annual production will determine project life. Dilution will extend project life and while all the *in situ* metal present will be recovered, it will take longer. Then, due to the impact of the time value of money, the NPV will be reduced based simply on the delayed benefit of the revenue stream. There will also be larger cumulative OpCosts albeit their impact will be ameliorated by the same low present value factors.

When poor metal reconciliation takes place the mine manager must have the wisdom and experience to recognise what is due to mathematical uncertainty and what is due to technical shortcomings. Lack of reconciliation may well fall within a defined envelope of uncertainty but people make mistakes and systems may need to be improved or re-engineered.

9.3 Management

Recent events in the South African mining industry have all the hallmarks of a major change in the political and commercial landscape. The concept of "cheap" labour-intensive mining was always an illusion. The Rustenburg-centred platinum mining industry based on underground mining of the narrow Merensky and UG2 horizons is going to have to mechanise, train up well-paid teams of skilled operators, provide decent family housing and reduce underground shift numbers by a significant factor. The problem in the Rustenburg area is the narrow reefs that are just over a metre wide: if the stopes are going to be wide enough to handle even low-profile mining equipment, dilution will increase and head grades will drop. Reductions in OpCosts per tonne need to offset loss of head grade if the strategy is to succeed. If mechanisation is implemented this will have social implications as unskilled labour makes way for skilled operators. These operational changes will inevitably draw attention to the potential of the Platreef (see Section 5.5.3) with high-grade zones which are at least 10 m wide and amenable to both open-pit and mechanised low-cost underground room and pillar mining operations.

The mining industry is now talking about "dis-economies of scale" which they consider to be a significant factor in mining's waning productivity. This is why there is a trend to spin off some of their operations into a separate company. If a company is too large it cannot be managed effectively. Scaling up mines just made them more complex to run while rapid staff turnover has left many mine managers without the people to monitor the increased complexity. There is also talk of how "silo mentality" has crept into management of a mining company as the large scale of the operations results in loss of perspective. Costs are being squeezed but may not be sufficient to make up for the fall in productivity.

9.4 Mining Methods

9.4.1 *Open-pit versus underground mining*

The advantages of open-pit mining are that it is safer than underground mining; it is associated with lower operating and Capex per unit mined and higher productivities are achievable; there is less delay in starting mining in an open pit; considerable flexibility that can also be exercised in the design of an open pit as is illustrated in Figure 9.3. Not only does this take into account the relationship between topography and stages of development in relation to the orientation of the ore body as illustrated in Figure 9.3 but also the sequence of back-stripping can be adjusted to optimise NPV. This is the basis for sophisticated pit design software.

The detailed relationship between blast-hole pattern and bench dimensions is given in Figure 9.4 and as can be seen from the inset in Figure 9.3, the relationship between the height of the benches and the distance between them determines the basic geometry of an open pit and its ultimate pit slope.

The optimisation of pit design takes into account different geometric designs constrained by estimates of revenue assumptions in a financial model. For example, if an escalation in metal prices is

Figure 9.3. A typical pit cross-section — Phases I–III are designed using increasing stripping ratios, thus increasing the volume of ore mined economically (after C.T. Shaw). Inset of Platreef at Overysel South, showing stable pit slope, ramp, blast-hole drilling and loading.

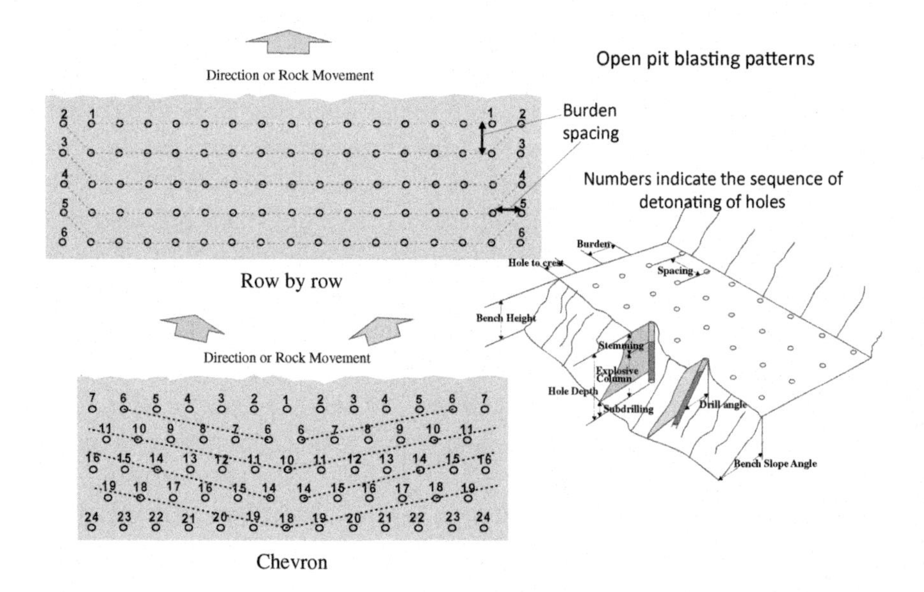

Figure 9.4. Open-pit geometry and terminology of benches (after C.T. Shaw).

assumed for the first 10 years this will not be the same as using real versus nominal values as outlined in Section 2.2. There will be no distortion of NPV, but it will probably give a different outcome when pit optimisation scenarios are considered. If you do not assume metal price escalation in the model then presumably you end up with a more conservative stripping ratio using identical scheduling. Assuming higher metal prices allows for a higher stripping ratio as the open pit deepens, you may be able to neutralise the effect through different scheduling involving moving more waste in the early years with a no metal pricing escalation scenario. If the metal prices do increase after the back strip is completed then there is a bonanza profit.

While traditionally higher outputs were possible from an open-pit compared to an underground mine, development using block-caving methods means that this is not necessarily always the case. In 2014 open-pit production at Chuquicamata, Chile, was 120,000 tonnes per day (tpd) (43.2 million tonnes per annum (Mtpa) of ore with the underground operation predicted to produce 140,000 tpd (50.4 Mtpa). This is a reversal of the usual scenario where the transition from surface to underground mining results in a drop-off in production. This is achievable because the block-caving mining method to be implemented will involve drawing from four separate levels simultaneously. The high rate of production from an underground operation is made possible because of the capacity of horizontal development compared to vertical shaft systems.

Palabora is now looking to develop Lift 2 in their block-caving operation. Daily total tonnage at Palabora from the open pit in 1989 was 215,000 tpd (77.4 Mtpa) with ore tonnage achieving 91,800 tpd (33.0 Mtpa) at a strip ratio of 1:1.25. The predicted production from the underground block-caving operation based on the shaft capacity was 25 Mtpa, although actual production reached a plateau at 12.5 Mtpa.

The importance of rock mechanics in determining the viability of block-cave projects became apparent. Multiple issues arising from the mechanical behaviour of the rock due to the alteration of the physical environment have a significant effect on the operation of block caves. Poor rock fragmentation has a large negative effect on the project value which arises owing to poor design of draw point layout and undercut size. This in turn results in a reduction of the extraction rate and an increase in the mining cost. In some cases, without appropriate geotechnical modelling at an early stage of project development, cave propagation may not occur. There have been cases where continued blasting is required, impacting on both the mining rate and mining cost.

When operating a block cave below an open pit the draw rate must constantly be monitored and the cave propagation modelled to avoid direct dilution of the orebody and, in the worst case, sidewall failure in the open pit which eventually contaminates the ore and causes dilution. This results in lower head grade and reduction of the mine life.

Crucially, the ramp-up duration for any block cave is determined by rock mechanics. Ramp-up requires a parabolic increase in production which in some cases can last over a decade. This would have a major impact and significantly reduce the financial returns generated by the project compared to original projections.

9.4.2 *Mine operating cost estimation*

As outlined in Section 8.3.2, there are two types of costs, Capex and OpCosts. Capex are costs in a particular year that will produce benefits in later years and thus may be depreciated over the life of the mine. This impacts on the NPV in a DCF model as tax relief in later years is discounted (see Section 2.3.1). A tax model such as loss carry forward would enhance NPV as relief is applied in the early stages of a project.

OpCosts are costs which only produce benefit for that year and can be used to offset tax liability in the year they are incurred. This means they provide greater benefit in a DCF than Capex. OpCosts are composed of direct costs which are directly attributable to production and indirect costs such as administration. They are also fixed or variable depending on output. In a gold processing plant cyanide will be consumed regardless of the grade of the ore but the cost of

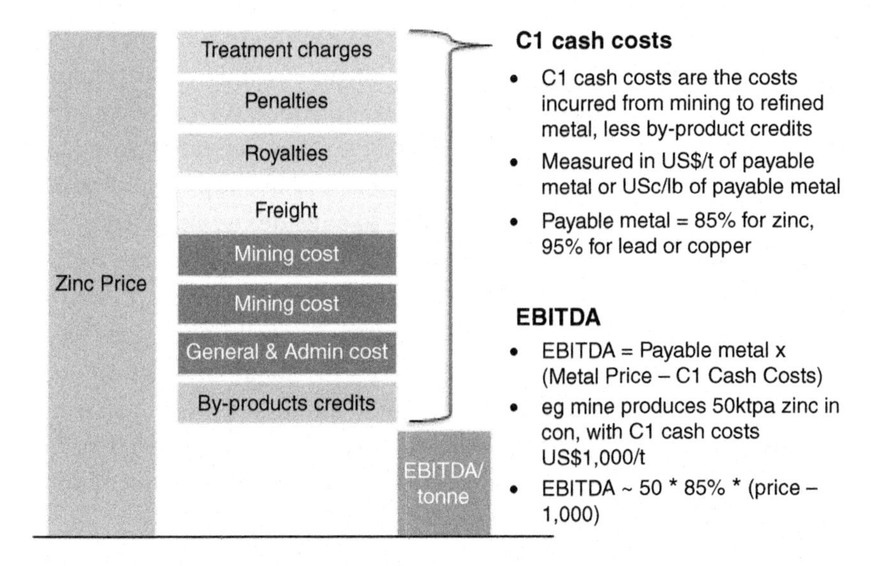

Figure 9.5. Break-down of cash costs.

crushing would be directly proportional to the rate of production. In operating mines costs are reported as outlined in Figure 9.5.

An insight into the process of cost estimation on a small open-pit gold resource is outlined below. The scenario assumes a small plant capacity of 500 tpd (around 150,000 tonnes per annum (tpa)). Entry-level mining normally starts at 250,000 tpa which at a grade of 10 g/tonne produces around 75,000 ounces assuming recovery of around 90%. A mid-sized operation would expect to produce 250,000 ounces per annum and head grade may be no more than 2 grams/tonne so the mine must produce 3.875 Mt of ore for processing. Waste rock movement may be five times this amount.

The actual process of mining comprises drilling, blasting, loading and hauling. The equipment required, performance and costs are given in Table 9.2. This provides the basis for generating the cost per tonne. It also demonstrates that the rate-determining stage needs to be identified and this is normally the loading process. Performance of the preceding drilling process and the following haulage stage requires a selection of equipment that balances capacity with flexibility.

The process of compiling the technical parameters needed for the scenario given in Table 9.2 might take 30 person days and is entirely inflexible. Any change in the basic assumptions on capacity effectively means that the whole exercise has to be repeated from first principles as quite different specifications for the equipment would be needed. When this process is expanded to consider a mid-sized operation, as described above, the amount of work involved in undertaking the detailed engineering increases proportionately and the inflexibility is further entrenched. Scenario analysis is really only possible at the pre-feasibility stage and not at the FTFS stage. Small-scale mining of the type illustrated is not usually viable not only because of high unit costs but because the number of operators needed to run a small operation may be no fewer than that needed for a mid-sized operation.

9.5 Coal Mining

9.5.1 *Strip mining*

Strip mining (or opencast mining) is a surface method applied almost uniquely to the coal-mining industry. The principal feature of the method is surface land reclamation contemporaneous with overburden stripping and haulage. Rather than being transported to waste dumps for disposal as in open-pit operations, overburden is cast directly into adjacent mined-out panels. Excavation and haulage is generally combined in one-unit operation (casting) and performed by a single machine. Uncovered coal is then mined by the conventional loader-hauler methods applied in open-pit mining.

Strip mining is a large-scale extraction technique, resulting in the lowest unit mining costs of any of the broadly used methods, high productivity, high recovery and greater safety records compared with underground mining.

Though expensive to purchase, the use of boom-type excavators such as draglines gives strip mining a major advantage over open-pit methods — the haulage (of overburden) component of the operating cycle is eliminated and contemporaneous surface rehabilitation is permitted. This results in more rapid uncovering of coal (and thus

Table 9.2. Cost estimates for benching.

Parameters		
Strip Ratio	Waste:Ore	4
Plant Capacity	t/day	500
Waste Movement	t/day	2,000
Total Movement	t/day	2,500
Rock Density	t/m^3	2.75
Bench Height	m	7

Unit Cost Summary		Ore	Waste
Drilling	$/t	0.43	0.20
Blasting	$/t	0.38	0.18
Loading	$/t	0.29	0.29
Hauling	$/t	1.48	0.23
Total	$/t	2.57	0.89

Drilling		Ore	Waste	Total
Burden	m	2.0	3.0	
Spacing	m	2.3	3.5	
Tonnes per Hole	t	88.6	199.2	
Sub Drill	m	0.6	0.9	
Metres per Hole	m	7.6	7.9	
Tonnes per Metre	t/m	11.7	25.2	
Metres per Day	m/day	43	79	122
Holes per Day	no./day	28	10	38
ROP	m/hr	20	20	
Drilling Hours per Day	hr/day	2.1	4.0	6
Working Hours per Day	no.	10	10	
Drills Required	no.	0.2	0.4	0.6

Drilling Costs		Ore	Waste
Bit	$/m	3	3
Rod	$/m	0.3	0.3
Stabiliser/ Hammer	$/m	0.2	0.2
Rig	$/hr	30	30
Cost per Hole	$/hole	38.0	39.5
Cost per Tonne	$/t	0.43	0.20
Cost per Hour	$/hr	100	100

Blasting		Ore	Waste
Hole Diameter	mm	108	108
Hole Volume	m^3	0.070	0.072
Stemming	m	2.0	2.0
Charge Length	m	5.6	5.9
Explosive Density	t/m^3	0.8	0.8
Charge Volume	m^3	0.051	0.054
Explosives per Hole	kg/hole	41	43
Powder Factor	kg/t	0.46	0.22

Blasting Costs		Ore	Waste
Explosives	$/kg	0.6	0.6
Detonator	$/hole	3	3
Booster	$/hole	3	3
Detonator Cord	$/hole	1	1
Delays	$/hole	2	2
Cost per Hole	$/hole	33.6	34.9
Cost per Tonne	$/hr	0.38	0.18

(*Continued*)

Table 9.2. (*Continued*)

Loading		Ore	Waste	Total
Bucket Capacity	m^3	2.5	2.5	
Fill Factor	%	0.91	0.91	
Cycle Time	min	1	1	
Cycles per Hour	no.	60	60	
Productivity	%	66	66	
Loose Density (70%)	t/m^3	1.93	1.93	
Rate of Production	t/hr	173	173	
Loading Hours per Day	hrs	2.9	11.5	
Working Hours per Day	no.	10	10	
Loaders Required	no.	0.3	1.2	1.4

Loading Costs		Ore	Waste
Cost per Hour	\$/hr	50	50
Cost per Tonne	\$/t	0.29	0.29

Hauling		Ore	Waste	Total
Capacity	m^3	12	12	
Cycle Time	min	45	7	
Cycles per Hour	no.	1.3	8.6	
Productivity	%	66	66	
Rate of Production	t/hr	20	131	
Working Hours per Day	hrs	10	10	
Trucks Required	no.	2.5	1.5	4.0

Hauling Costs		Ore	Waste
Cost per Hour	\$/hr	30	30
Cost per Tonne	\$/t	1.48	0.23

greater production levels) and lower unit OpCosts than a similar-sized open-pit operation.

Depth constraints are the major limitation of strip mining, 100 m being the approximate technological and economic limit of the method. This translates to a maximum economic stripping ratio of approximately 40:1. The nature of the cyclic casting operation means that strip mining is a fairly inflexible mining method and is only feasible in the extraction of relatively flat-lying, regular coal seams. The initial capital investment associated with equipment purchase is considerable, though this is compensated for by reduced OpCosts as governed by the economics of scale.

9.5.2 *Longwall mining*

A series of longwall panels is laid out. Each longwall will be mined in turn with a narrow pillar left between it and the longwall next to it. The layout of the longwalls is dependent on the geology of the coal seams, the geology of the overlying rock strata, the geotechnical characteristics of the coal and the surrounding rock, any dislocations such as faults or folds affecting the seam to be mined, and any restrictions in force regarding the undermining of whatever is on the surface above the coal to be mined. Any unexpected faults in particular can severely affect the operation of a longwall — if the throw on the fault is large enough the whole longwall may have to be redeveloped.

The longwall shearer (or in some longwalls, a plough) cuts the coal from a face of 1,000 feet (300 m) or more between gate roads at each end of the face. In European practice the gate road will be a single tunnel. In American practice the gate roads are multiple tunnels mined on a room-and-pillar basis. On the longwall the loosened coal falls onto an armoured face conveyor that takes the coal to another conveyor in the gate road. Longwall systems have their own hydraulic roof supports, called chocks or shields, which advance with the machine as mining progresses.

If the gate roads have been developed before starting the longwall and the longwall is mined back between these established roads, then it is called "retreat mining". If on the other hand the gate roads are developed alongside the longwall as it progresses forward, then it is called "advance mining".

As the longwall mining equipment mines forward, the overlying rock that is no longer supported by coal must cave behind the line of shields in a controlled manner. This caving is essential for the stability of the system.

This is a highly productive underground mining method and is relatively safe as the operators at all times work under the protection of the powered roof supports.

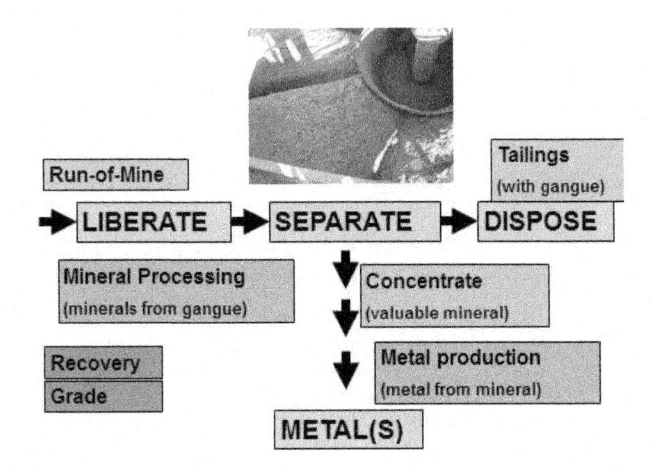

Figure 9.6. General process flowsheet and terminology (after J. Cilliers).

9.6 Extractive Metallurgy

9.6.1 *Mineral processing*

It is said that mineral extraction starts at the stage after rock comes up the conveyor belt and into the processing plant. It is certainly true that a failure to understand the complexity of ores and to incorporate appropriate recognition into the design of the plant is a significant source of technical risk. The key stages are given in Figure 9.6.

The key function of mineral processing is to achieve an optimum balance between what you want and what you don't want. Minerals are liberated using comminution, i.e. crushing and grinding, and this is a very energy-dependent stage which accounts for around 40% of the total energy used in mineral processing operations. It also accounts for 4% of the world's total electrical energy consumption, source:

http://eex.gov.au/2013/11/crushing-energy-costs-in-the-mining-sector/.

While pre-breaking takes place during blasting at the mining stage, explosives are a mixture of ammonium nitrate and fuel oil (ANFO) and so there is an energy cost at this stage. Over-fragmentation during blasting complicates grade control and is less

Figure 9.7. Flotation cells at DRD's Ergo gold operation. Reproduced with their permission.

efficient than that which can be achieved using primary crushing. Run of mine (ROM) ore is crushed to a fine particle size and mixed with water to form a slurry, which is fed into the grinding mill where it is reduced to a very fine pulp.

Mineral separation uses differences in mineral properties. With flotation the pulp is mixed with reagents (usually xanthate and a suitable frother) and agitated with compressed air. Air bubbles form and collect the hydrophobic sulphide particles. The sulphide-loaded air bubbles float to the surface and concentrate as a froth which is then collected (see Figure 9.7). Flotation usually occurs as several stages, increasing the metal grade of the concentrate as it passes through each stage. Polymetallic sulphides can be recovered from the residual pulp by using suitable alternative reagents.

Inevitably, some of the metal in the initial ore feed is never recovered and around 90% recovery rate is the best that can realistically be expected. However, recovery rates may fall further, owing to attempts

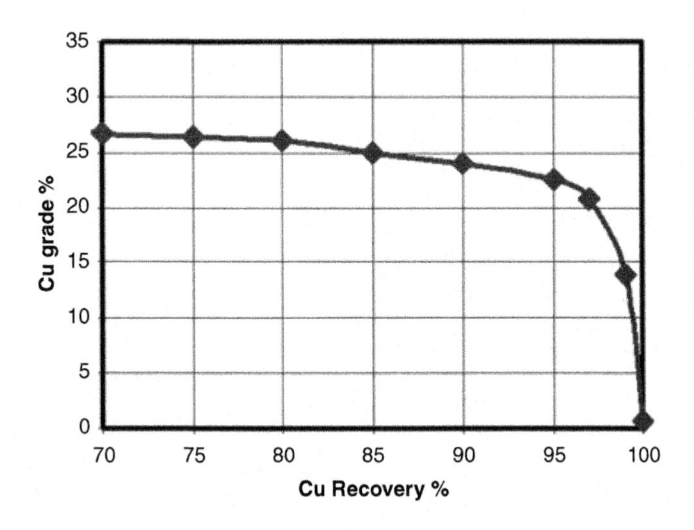

Figure 9.8. Grade–recovery relationship (after J. Cilliers).

to produce a higher grade concentrate — 60–80% nickel metal recovery is not uncommon. In contrast, most copper by-products will normally be recoverable at higher rates.

For a copper scenario in which the plant is producing a copper concentrate the higher the recovery the lower will be the copper content in the concentrate. This is illustrated in Figure 9.8.

As can be seen, if the plant was operated such that 97% of the copper was recovered, the copper grade in the concentrate would be 20%. A feed of 1,000 tonnes of ore at a grade of 0.6% copper present as chalcopyrite ($CuFeS_2$), which comprises around 35% copper, would contain 17.1 tonnes of the mineral in the feed. Of the 6,000 kg of copper processed, 5,820 kg would be recovered (16.63 tonnes of chalcopyrite). Each tonne of concentrate, assuming it is made up entirely of chalcopyrite, would at a copper price of $7,500/tonne be worth $2,625. Transport and shipping costs would therefore be a small proportion of the overall contained value, allowing producers to enter into off-take agreements with smelters (see Section 9.5.2).

If recovery was reduced to 85% then the copper grade in the concentrate would increase to 25%, but the overall copper produced would drop to 5,100 kg. As revenue is directly dependent on copper produced there is clearly benefit from operating at the higher recovery rate. There are, however, diminishing returns on pushing recovery further as copper content in the concentrate would drop steeply.Transport costs increase for every unit of copper and as the copper content of the concentrate fell then the smelters would start imposing a penalty. Some smelters may reject the product completely if the grade is too low. At a 100% recovery, of course the grade of the "concentrate" is the same as the feed.

As some of the metal in the initial ore feed is never recovered and ends up with the waste material, usually in the tailings, controlled disposal is a key part of any mining operation. Residual sulphide minerals present in the tailings will oxidise in the presence of water and air resulting in acid mine drainage (AMD). The common mineral is pyrite that will react as follows:

$$4FeS_2 + 22O_2 + 14\ H_2 \rightarrow 4Fe(OH)_3(solid) + 8SO_4^{2-}(aq) + 16\ H^+(aq).$$

Iron oxides will be precipitated, giving the characteristic red colour to AMD while the high acidity means that other metal species will be dissolved and precipitated. The reaction can be halted by removing the presence of either water or oxygen, and this is achieved by initially keeping the tailings under water in an anaerobic environment and covering with soil after mining is completed. Long-term monitoring will be required.

A raised embankment is the most common construction where tailings are put through a cyclone with the slime proportion discharged to the centre of the impoundment and the sand fraction to the beach behind the crest (see Figure 9.9).

9.6.2 *Extractive metallurgy*

Once a concentrate has been produced the contained metals need to be extracted appropriately using hydro- and pyro-metallurgical techniques. This is illustrated in Figure 9.10.

In the case of nickel, metal concentrate is filtered and dried to a fine powder in preparation for the smelter. Smelters are not usually located on the mine site and may accept concentrate from a variety of individual operations. The price paid by the smelter for the

Figure 9.9. Wall of tailings dam from historical mining in the East Rand Witwatersrand gold-mining operation undergoing reclamation with the slurry being reprocessed through DRD's Ergo gold operation before being discharged onto a new tailings dam. This type of operation is rehabilitating land that can be utilised for industrial development. Reproduced with their permission.

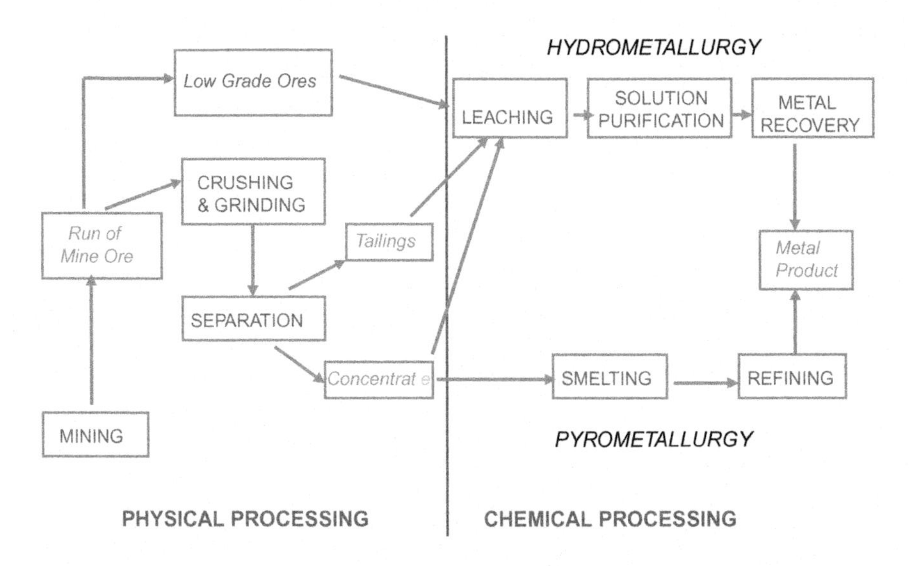

Figure 9.10. Process routes for metalliferous ores (after J. Cilliers).

concentrate will be based on an off-take agreement. Arrangements entered into include net smelter returns (NSR) and toll smelting and toll refining. A NSR might be as low as 60% of prevailing spot price and would incorporate penalties when concentrates do not meet pre-determined specifications. An off-take agreement is usually a confidential contract and it is often best to assume NSR and not spot prices in financial models.

The concentrate is exposed to a process known as flash smelting, which utilises energy from the exothermic reactions occurring when the iron and sulphur (contained in the powder) are burnt in an oxidising atmosphere The heavier molten metal sulphide sinks to the floor of the smelter while the lighter silicate melt floats to the top forming a slag.

The nickel matte is recovered and dried in preparation for further refining. Once again, some nickel is lost in the smelting process due to entrainment of the matte and as dissolved NiO. In general, the higher the matte grades, the higher the nickel losses. The slag is fed back into the smelter in an attempt to maximise nickel recovery. Flash smelting typically produces a granulated nickel matte containing approximately 75% nickel.

In South Africa low sulphide PGE UG2 concentrate is blended with high sulphide PGE Merensky Reef concentrate (see Section 9.5). to permit conventional (1,400°C) smelting. This means that the highly refractory chromite with a melting point of 2,200°C does not build up in the furnace. Chromium (Cr^{3+}) essentially partitions into the slag. The phase chemical systems and thermodynamic controls are identical to those outlined in Sections 5.5.2 and 5.5.3. The metallurgists are simply re-creating the conditions present in nature during the crystallisation of the Bushveld. In order to process a pure low sulphide UG2 concentrate then enhanced chromium partitioning can be achieved by operating the furnace at a higher temperature, but this requires a design that incorporates graphite/copper cooling when operated at a temperature of 1,600°C. This system obviously provides a commercial advantage where an increasing proportion of the PGE comes from the UG2 rather than the Merensky Reef. The high sulphide content of the Platreef provides a potential additional operational advantage as it ensures optimal proportions of matte and slag in the charge.

While PGE projects are a niche business, treatment of the extractive metallurgical issues is illustrative of the core theme for the "metals" part of this book. It is important that in evaluating a project the assessor looks beyond a product that comprises a concentrate of sulphide or oxide minerals to the associated contained metals. These can be directly linked to spot prices which allows for greater sophistication in financial modelling. NSRs usually imply fixed revenue assumptions, but if metal is produced as a final product and sold into a commodities market then the treatment of price volatility can be modelled using the principles of quantitative finance. PGE projects also demonstrates how financial modelling needed for an investment decisions in down-stream processing must be constrained by the thermodynamics of smelting.

Refining facilities may be located at a distance from the mine and smelter as they have considerable power requirements and so seek to locate in a region where power is both readily available and cheap. Nickel refineries employ a technique known as the Sherritt–Gordon process in which the nickel matte slurry is treated to remove selectively unwanted metals, sulphur and finally copper. The resulting solution is then purified in a nickel reduction process, leaving a high-grade nickel metal product. Nickel is usually produced as marketable briquettes with a 99.8% Ni purity. Downstream refining can be considered a separate business unit.

The conclusion for many investors is that it is best to stick to gold as the operation will produce bullion on site. The only remaining stage is refining and a high grade non-refractory gold deposit may well be associated with low technical risk. If that logic is followed through then diamonds are an even better option as all that is being produced on site is the raw mineral. The very low grades associated with primary deposits of kimberlite or alluvial gravels plus the challenges of determining price per carat, however, raises separate considerations that make these some of the most difficult projects to assess.

9.6.3 *Coal beneficiation*

Coal beneficiation is the stage of production when ROM coal is processed into a range of clean, graded and uniform products suitable for the commercial market (see Sections 8.7.1 and 10.7). Following extraction, coal usually requires crushing to reduce it to a usable and consistent size, while washing may be necessary to remove pieces of rock or mineral that are present and so improve the overall quality of the product. If ROM coal is of variable grade, blending may be required to ensure a consistent quality of coal to the end-user.

The extent of beneficiation carried out is dependent on customer requirements and the physical characteristics of ROM material. In some cases ROM coal is of such quality that it meets user specification without the need for any treatment, in which case it is merely crushed and screened to deliver the specified product size. The capital requirements for developing a processing plant are generally low compared with the expenditure on mining equipment.

9.7 Capex Estimation

A typical Capex estimation at the PFS stage for a project that comprises an integrated mining and processing capability will normally also include provision for infrastructure. Capex estimates will be associated with an inherent uncertainty of ±20%.

Funding requirements at the FTFS stage must incorporate the cost of the EPC&M contract (see Section 8.5.3) which can be up to 15% of the total Capex. There is also the level of contingency that should be included as part of the overall funding requirement. No promotor wants to run out of money before the project is completed. This risk is compounded when the promotors insist that the independent engineers revise down Capex estimates. This fails to fully appreciate that the final costs under an EPC&M contract will only be revealed once tenders have been received. If initial estimates have been unrealistic then the inevitable budget "blow-out" will occur during construction as reality sets in.

Unrealistic Capex estimates will probably be associated with inadequate provision for contingency. All construction stops and the operation fails to get into production on schedule and the start of revenue is delayed. The whole project then becomes fatally flawed, as the NPV on which the investment decision was made is never achieved.

The pragmatic approach is to accept that even at the FTFS stage Capex estimates will be associated with an inherent uncertainty of ±10% (see Section 8.5.3). Funding requirements should then assume that the final Capex would be the base case plus 10%. Contingency is then treated as covering the unexpected such as pre-production development encountering an aquifer which requires unprecedented pumping capacity. It should not be used to accommodate changes to the original design imposed by the promotors on the engineers implementing the EPC&M contract. Within that context, a contingency of 10% on top of the base case plus 10% should be adequate.

Contingency can be as high as 20% but it is seldom clear how this is determined. If it is on top of the upper cost estimate that will be expensive if funded in part from debt. It also increases the challenge of securing the balance needed from the equity component.

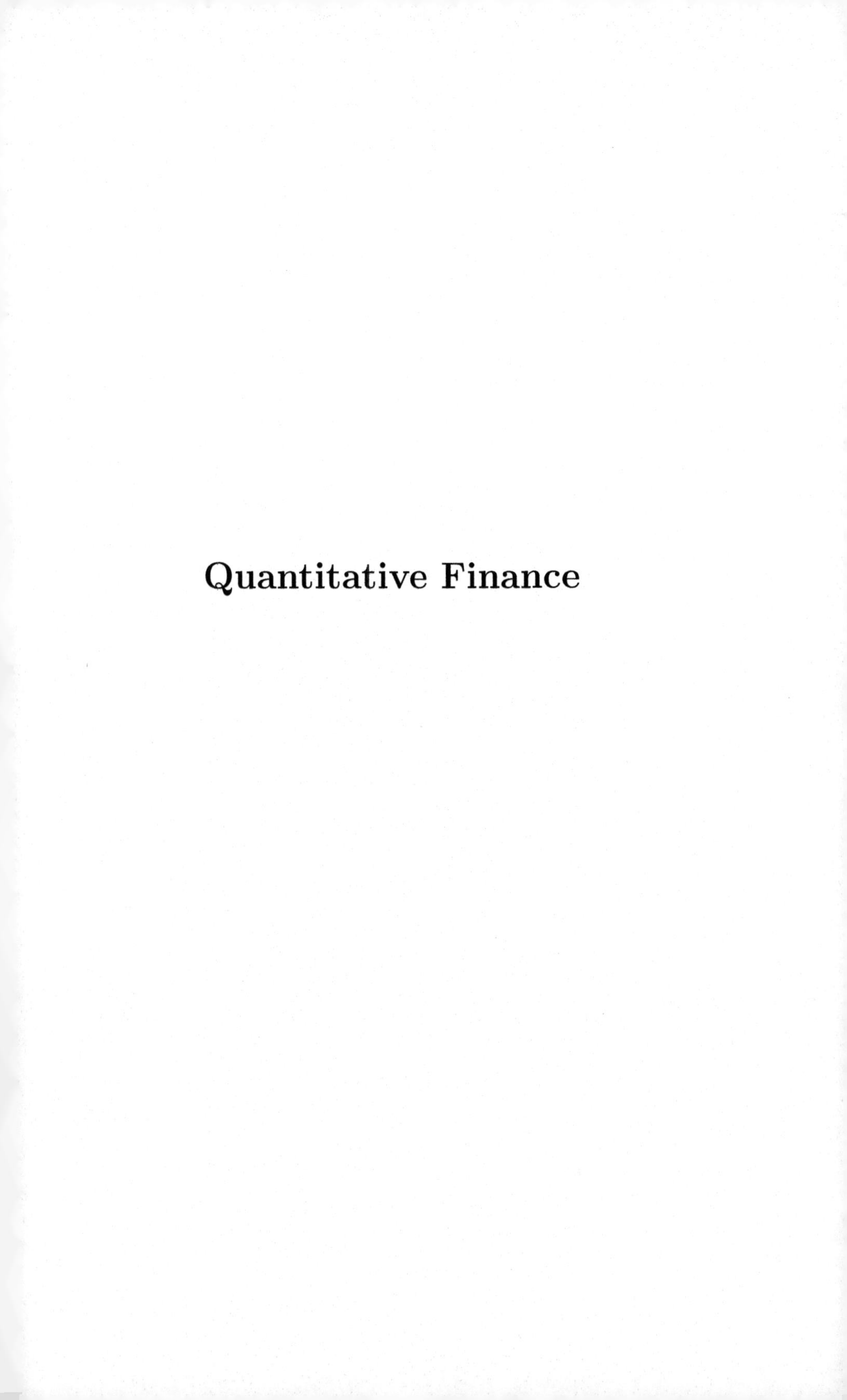

Quantitative Finance

Chapter 10

Quantitative Finance and Financial Engineering

10.1 Introduction

Once the resource estimation and conventional engineering stages have been completed financial analysis can be undertaken using the tools of quantitative finance. The inherent uncertainties present in technical assumptions have their analogues in the probabilistic approach used in investment and portfolio management and the pricing of a range of financial derivatives. These are considered in this chapter.

10.2 Decision Tree

Once NPV optimisation at the pre-feasibility stage has been completed as explained in Section 8.3.3 commercial decisions need to be taken. This process is aided by the use of a decision tree as illustrated in Figure 10.1 starting from a MC simulation of the primary technical variables outlined in Section 2.4. This is based on simultaneous treatment of grade, plant recovery, OpCost and Capex at the PFS stage (see Section 8.3.3). We select these variables as they do not change project life and therefore the base case model. Capacity and size of the reserve would change project life and therefore the base case model. A MC simulation on these variables would not be valid.

In undertaking a MC simulation to determine the robustness of a project there is no merit in flexing metal or commodity price as this generates a self-evident outcome (higher price will always improve

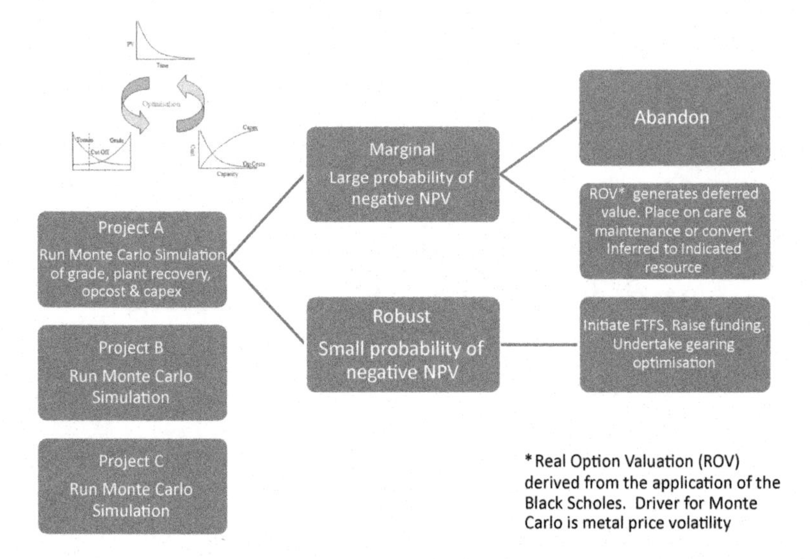

Figure 10.1. Decision tree at pre-feasibility stage after NPV optimisation with NPV optimisation derived from Figure 8.2.

the NPV). The best way to take into account the influence of metal prices is through the technique of Real Option Valuation (ROV).

10.3 Real Option Valuation

As outlined in Section 4.8 the application of ROV can be linked to practical considerations such as the duration of exploration licences. The steps that need to be followed in setting up a BS option pricing model in undertaking a ROV of a gold project based on deferring the development in the expectation of more favourable metal prices involve the following:

- Determine the PV of the free cash flows during the operation period based on a cash flow model discounted by the CAPM (see Section 2.3.2) after removing the effect of financial gearing.
- Determine the strike price which is the PV of the development costs.
- Estimate the volatility of the PV of the free cash flow to the firm (FCFF) using as the driver to a MC simulation the forward metal

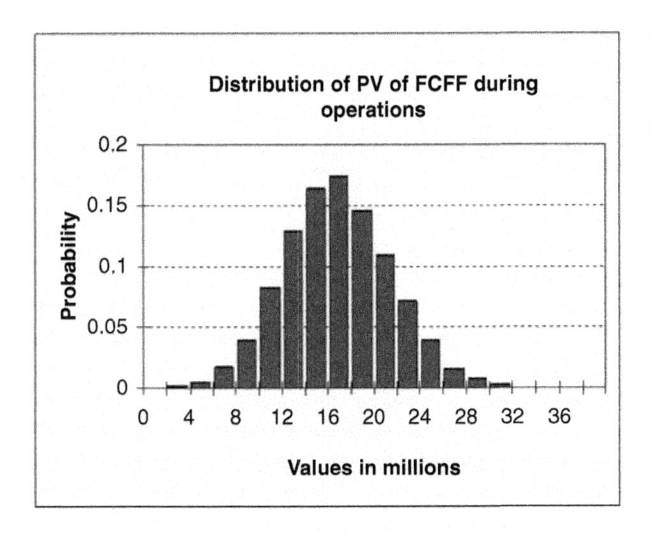

Figure 10.2. Output from a MC simulation.

prices and their volatility. The MC simulation will yield the volatility of the base case value (i.e. the PV of FCFF during the operating period).

- Time to expiration which is the time period in which to start development and could, for example, be the period after which a licence is either renewed or allowed to lapse.
- The risk-free rate.

The mean base case value from the output from the MC simulation will give the PV of the FCFF and the standard deviation (the volatility of the project) as shown in Figure 10.2.

The values are applied to the BS option pricing model as outlined in Figure 10.3. If this is treated as a European option which can only be exercised on the expiry date then it is relatively simple to value. If this is significantly higher than the PV then it is worth holding onto the licences. If not then the licence can be returned, as illustrated in Figure 10.1.

The techniques of ROV can be applied to a range of different natural resource projects including oil and gas. Scenario analysis of an Athabasca Oil Sands Financial Model Steam-Assisted Gravity Drainage (see Section 6.4) is illustrated in Figure 10.3 based on an

Employ the Black & Scholes option pricing model, which is based on:

- The Base Case value — (PV of FCFF during the operational period)
- The strike price — (PV of the development costs)
- Volatility — (of the PV of FCFF during the operational period)
- Time to expiration — (time after which the Project is developed i.e. 3 years)
- The risk-free rate — (the yield on 3-year sovereign debt)

The Black & Scholes equation for a Call Option:

$$V = N(d_1)\,A - N(d_2)\,Xe^{-rT}$$

Can be interpreted as follows:

$N(d_1)\,A$	the expected value of A if A > X at Expiration
$N(d_2)$	risk-neutral probability of A > X at Expiration
Xe^{-rT}	the Present Value of the development costs

Where:

- V — current value of Call Option
- A — PV of the FCFF during the operational period
- X — cost of development
- r — risk-free rate of return
- T — time to expiration
- σ — volatility of the PV of the FCFF during the operational period
- $N(d_1)$ & $N(d_2)$ — are the values of the normal distribution at d_1 and d_2
- d_1 — $[\ln(A/X) + (r + 0.5\,\sigma^2)\,T\}\,/\,\sigma T^{0.5}$
- d_2 — $d_1 - \sigma T_{0.5}$

Figure 10.3. Application of the BS option pricing model to a hypothetical gold project. Reproduced with the agreement of Chris Worcester.

unpublished dissertation by Bernard Guerbet, Imperial College London in 2009 when oil prices were low and this type of project was considered to be commercially marginal (see Figure 10.4). The scenario is based on the extraction of bitumen at a depth between 400 and 800 m below the surface. Each barrel of extracted bitumen will be upgraded into a barrel of synthetic crude oil (see Section 6.3).

The start of the Phase 1 construction period of 3 years was scheduled for the period 2009–2011. First oil production was in 2012, first peak oil production of 25,000 barrels per day (bpd) in 2015. The start of Phase 2 was scheduled for 2015 with an increase in production capacity from 2015 to 2022. Phase 2 was to be completed in 2019

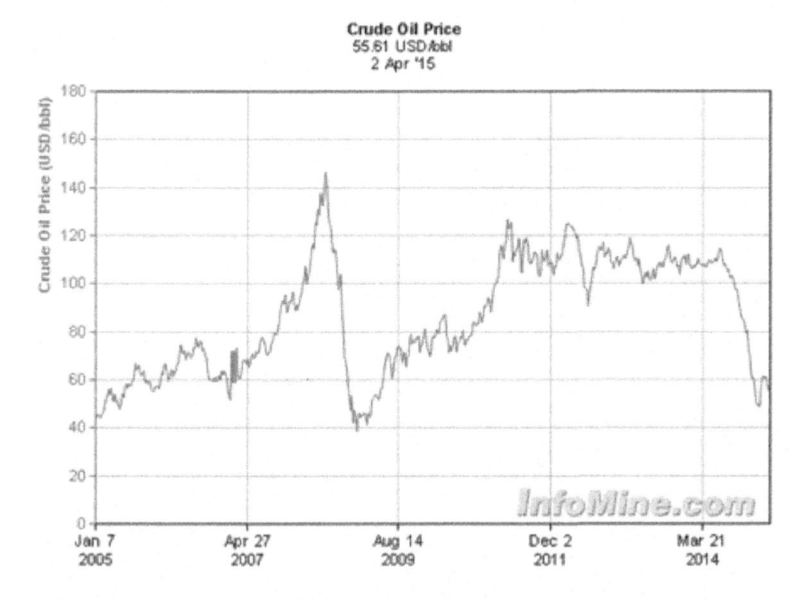

Figure 10.4. Crude oil price. Reproduced with the permission of InfoMine.

with final peak oil production of 100,000 bpd in 2022. The plateau was reached from 2022 to 2045. A decline of production started in 2046 with end of production in late 2049. The results of the technique of ROV are given in Figure 10.5.

As can be seen, the model demonstrated that deferring the development to 2012 would have coincided with the recovery of the oil price from \$70 in 2009 to \$110 in 2012 notwithstanding that a simple DCF model demonstrates declining NPV over the same period. The correct decision would have been to retain the right to develop the project but defer the decision to invest.

10.4 Funding

10.4.1 *Tax and company structure*

The framework for a JV agreement is outlined in Chapter 3 (Figure 3.6) and illustrates how the local JV company holds the mineral licence. This would obviously be geographically constrained but funding may well be sourced externally. It is not unusual for exploration

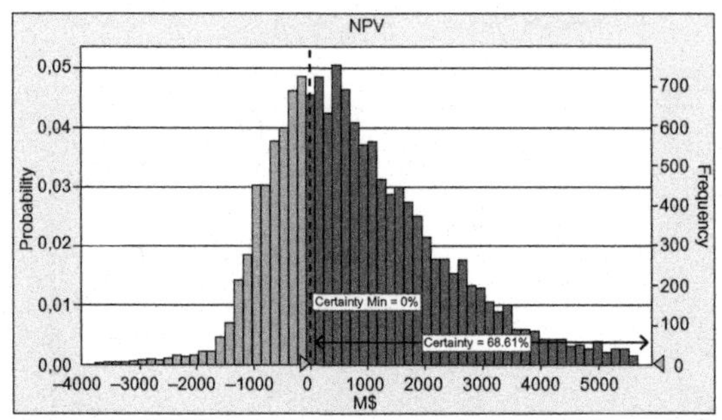

Real Option Valuation

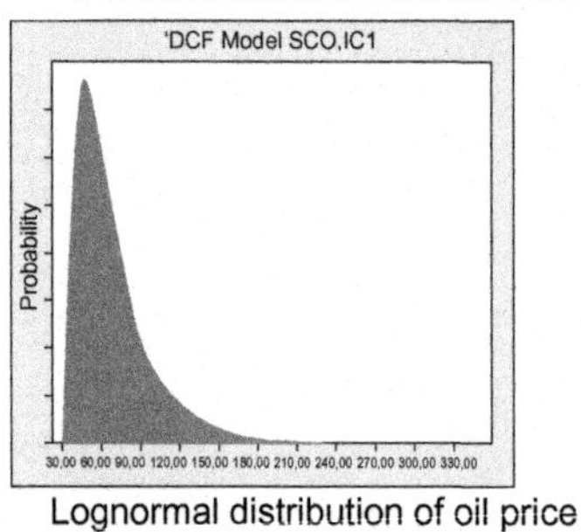

Lognormal distribution of oil price
Mean of $71, SD of 37.7

Time Delay (y)	Real Option Value (M$)	NPV (M$)
0	1,005	1,005
2	1,147	898
3	1,203	716
5	1,096	570

Figure 10.5. Diagram showing the distribution of NPV derived from the volatility of oil price and the ROVs generated from the BS model.

to be funded by a high net worth individual via a private placement. Privacy issues can dictate that the holding company may be registered in a tax haven and it is only when the project starts showing potential and follow-up stages of funding a development are considered that the tax implications of the set-up arrangements emerge. Unless there is a dual tax agreement between the country in which the project is located and the country where the company that will hold the asset is to be hosted, then there will be draconian withholding taxes.

These macro considerations would eliminate relatively subtle advantages arising from the tax relief on interest paid on debt and treatment of depreciation. This restructuring can be complex and IFRS need to be applied to local JV accounts which may have to be re-audited. Furthermore, there could be a need to prepare close-out accounts for the offshore vehicle through which the original

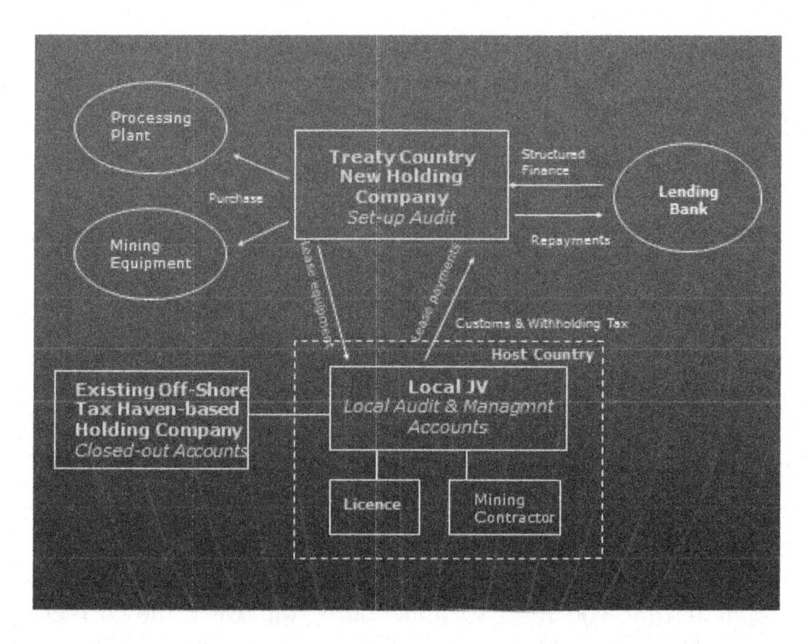

Figure 10.6. Company structure for a typical medium-scale mining project.

exploration funding was provided. Then there would be the need for a set-up audit for the new company. This then allows more sophisticated arrangements to be put in place relating to leasing and payment of VAT, as illustrated in Figure 10.6. While it is obvious that it would be best if the most tax-efficient company structure was put in place at the exploration stage, strategic considerations often take precedence (see Section 3.8.3).

Mineral taxation regimes can be complex and each country will have a different regime. These are sometimes derived from petroleum fiscal models so some countries impose an excess profit tax (EPT) common in petroleum projects. It applies once the cash flows on the project generate a rate of return that exceeds a specified threshold. After that the additional tax is charged on a sliding scale up to a defined upper threshold. EPT has a disproportionate impact on small projects where investors are seeking a high IRR on their investment rather than a significant NPV, as illustrated in Table 2.1.

A commercial discovery bonus is common in the petroleum industry where it is perhaps more obvious when this event occurs (a single well can start producing oil, but a single drill hole into mineralisation does not convert this into a mineral project), but the only justification for applying this tax to mineral projects is where a company has benefitted from state-funded geological surveys.

The royalty rate is a tax applied directly to revenue, before the OpCosts are taken into consideration. It is therefore not possible to avoid and is essentially a resource rent that recognises that extraction of a natural resource permanently removes future value.

10.4.2 *Gearing optimisation*

Once you decide to proceed to the FTFS stage (see Section 8.5.2) then you need to consider funding options. This requires building a financial model which incorporates both debt and equity, and then discounting the cash flows using WACC (see Section 2.5.2). The starting point is that you need to make sure you select the correct net cash flow row in the financial model spreadsheet. That is NPV ATBI (after normal tax that includes relief on OpCosts but before tax relief on interest). DCFs for PAIBT and PAIT inflate the NPV.

The more general point is that when funding includes a component of debt, DCFs need to be integrated into the financial accounts otherwise you end up with an incorrect NPV. As WACC will be a function of the cost of equity derived from the CAPM and the variable cost of debt, further optimisation of NPV can be achieved through gearing.

In reality corporate cost of capital is what is imposed on subsidiaries of the major mining companies much to the frustration of project teams on an operating mine seeking to justify capital expenditure. They are obliged to use a high cost of capital that in many cases artificially depresses the NPV and detracts from expansion plans. The disinvestment of a major from a former subsidiary may well permit the new owners to go directly to the investment banks and raise debt independently. This can have a significant positive impact on the viability of the project.

The investment bank can also assist with an equity rights issue for an independent operation as well as being able to set up matching

Table 10.1. Base case for an open-pit nickel mine.

Mine reserves	150.4 million tonnes
Commodity grade/plant recovery	Nickel 0.6% (65% recovery)
	Copper 0.1% (90% recovery)
Ore SG	3.2
Mining capacity	30,952 tonnes ore per day
Stripping ratio	3.2:1
Mining recovery	85%
Dilution	5%
Working days per year	360
Commodity sale prices	Nickel $14000/t
	Copper $3000/t
Capex mine/mill	$165 million
Capex processing plant	$275 million
Capex overheads	10%
Permitting years	1
Construction years	2 (60% Capex yr 1, 40% yr 2)
OpCosts	$2.5/t mining ore
	$0.75/t mining waste
	$3.5/t processing
	$10 million per year fixed costs
Working capital	25% of annual OpCosts

hedges to ensure the promoters can cover repayment of interest, balance of loan and OpCosts. The proportion of metal that needs to be hedged is a constraint on the level of gearing as equity investors want the project to benefit from increases in prices.

10.5 Base Metals Case History

The critical issues in the project financing of nickel sulphide projects can now be assessed by modelling a hypothetical case scenario set in Western Australia. Though not based on any existing operations, the case example is a realistic representation of current nickel mining practice in Western Australia, and the figures used are reasonable reflections of those that might be encountered.

The project sponsors are seeking up to 70% debt funding of the total Capex and have been offered an interest rate of 8.5% on the loan capital. The assumed cost of equity is 15%. NPV will be calculated using the WACC, reverting to the cost of equity once debt has been paid off. A tax rate of 30% is assumed.

The banks require straight loan annual repayments over 5 years and are willing to grant a 2-year grace period on repayments. An upfront fee of 1% of the total debt is assumed with a 1% commitment fee. Fixed financing charges will amount to US $0.5 million and there is a contingency of 10% to act as a cushion against unexpected cost rises. The cash flows for the first 5 years are given in Table 10.2.

The annual cover ratios as outlined in Section 8.6 are of particular interest to the lender as they provide an indication of the project's ability to function economically while servicing debt. Generally lenders will require that the interest cover, principal cover and debt service cover ratios in any given production year are considerably greater than 1 in order to satisfy concerns over the project defaulting on the debt repayment schedule.

Gearing optimisation based on different levels of debt are summarised in Table 10.4. This demonstrates that optimisation of NPV is achieved at a traditional debt:equity ratio of 60:40.

Any detailed review of the model also needs to take into account metal price assumptions and hedging. At a 60:40 debt:equity ratio some 30% of nickel will need to be hedged to cover OpCosts and debt service repayments. While this can detract somewhat from the potential for the project to benefit from an increase in nickel price that equity investors would look for, they do have the comfort that in order to secure this level of debt the bankers would have undertaken a rigorous technical review of the project.

Capex and OpCosts assumptions will be derived from the FTFS and the uncertainty in Capex estimates (normally $\pm 10\%$) should not be confused with provision for contingency that is to cover the unexpected. It would be prudent to set up Capex funding requirements at the base case plus 10%. Contingency of 5% would be prudent, 10% would be expensive.

Reserve tail assumptions are based on the premise that once the debt has been repaid, at least 30% of the original reserve should remain as the lender's final recourse. While this may seem generous, technical risk needs to be considered, as indicated in Figure 10.7.

Table 10.2. Sample of DCF model worksheet — nickel sulphide project example.

WA_Ni_Sulphide	1	2	3	4	5	6	7	8	9	10
CashFlow										
Inflation	100%	100.00%	100.00%	100.00%	100.00%	100.00%	100.00%	100.00%	100.00%	100.00%
Reserves at Start Year (M/y)	127.84	127.84	127.84	127.84	121.47	112.98	102.37	91.76	81.15	70.53
Ore Tonnage Mined (M/y)	0.00	0.00	0.00	6.37	8.49	10.61	10.61	10.61	10.61	10.61
Rem Reserves at End Year (M/y)	134.23	127.84	127.84	121.47	112.98	102.37	91.76	81.15	70.53	59.92
Gross Rev (M$)	0.00	0.00	0.00	364.85	486.47	608.08	608.08	608.08	608.08	608.08
Royalties (M$)	0.00	0.00	0.00	0.00	−10.95	−14.59	−18.24	−18.24	−18.24	−18.24
Annual Op Costs (M$)	0.00	0.00	0.00	−00.16	−84.88	−103.60	−103.60	−103.60	−103.60	−103.60
Op Margin (M$)	*0.00*	*0.00*	*0.00*	*298.69*	*390.64*	*489.89*	*486.24*	*486.24*	*486.24*	*486.24*
Capex (M$)	0.00	−290.40	−193.60	0.00	0.00	0.00	0.00	0.00	0.00	0.00
Wcapt (M$)	0.00	0.00	0.00	−16.54	−21.22	−25.90	−25.90	−25.90	−25.90	−25.90
Change in Working Capital (M$)	0.00	0.00	0.00	−16.54	−4.68	−4.68	0.00	0.00	0.00	0.00
Environmental & Closure Costs (M$)	0.00	0.00	0.00	−1.50	−1.50	−1.50	−1.50	−1.50	−1.50	−1.50
CashFlow (Before Interest and Tax) (M$)	**0.00**	**−290.40**	**−193.60**	**280.65**	**384.46**	**483.71**	**484.74**	**484.74**	**484.74**	**484.74**
Ann Disc. CF (Before Interest and Tax) (M$)	0.00	*−240.00*	*−145.45*	*191.69*	*238.72*	*273.04*	*248.75*	*226.13*	*205.58*	*186.89*
Ace Cash Flow (Before Interest and Tax) (M$)	*0.00*	*−290.40*	*−484.00*	*−203.35*	*181.11*	*664.82*	*1,149.56*	*1,634.29*	*2,119.03*	*2,603.77*
Tax Paid Before Funding (M$)	0.00	0.00	0.00	−77.68	−105.26	−135.03	−133.94	−133.94	−133.94	−133.94

(Continued)

Table 10.2. *(Continued)*

WA_Ni_Sulphide		1	2	3	4	5	6	7	8	9	10
CashFlow (Net of Tax) (M$)		**0.00**	**−290.40**	**−193.60**	**202.97**	**279.20**	**348.67**	**350.80**	**350.80**	**350.80**	**350.80**
Ann Dis. CF (Net of Tax) (M$)		*0.00*	*−245.93*	*−150.88*	*145.57*	*184.27*	*206.68*	*183.25*	*156.33*	*125.28*	*86.71*
Accumulated Cash Balance (M$)		*0.00*	*−290.40*	*−484.00*	*−281.03*	*−1.83*	*346.85*	*697.64*	*1,048.44*	*1,399.24*	*1,750.04*
Total Costs Associated with Financing (M$)		−8.84	−1.74	0.00	0.00	0.00	0.00	0.00	0.00	0.00	0.00
Interest (M$)		−0.56	−20.92	−35.71	−35.71	−35.71	−32.14	−25.00	−17.86	−10.71	−3.57
Tax Paid After Funding (M$)		0.00	0.00	0.00	−66.96	−94.55	−125.39	−126.44	−128.58	−130.73	−132.87
Debt Funding (M$)		6.61	239.47	174.07	0.00	0.00	−84.03	−84.03	−84.03	−84.03	−84.03
Equity Funding (M$)		2.83	102.63	74.60	0.00	0.00	0.00	0.00	0.00	0.00	0.00
CashFlow (After Funding and Debt Service) (M$)		**0.05**	**29.04**	**19.36**	**177.97**	**254.20**	**242.14**	**249.27**	**254.27**	**259.27**	**264.27**
Cumulative CashFlow (After Funding and Debt Service) ((M$)		*0.05*	*29.09*	*48.45*	*226.42*	*480.62*	*722.77*	*972.04*	*1,226.30*	*1,485.57*	*1,749.84*
Pay Back Period (ATBI) (Years)	5.01										
Max Cash Exposure (ATBI) (M$)	−484.00										
NPV (ATBI)(M$):	1008.36										
IRR (ATBI):	46.19%										

Table 10.3. Financial ratios for nickel sulphide project example assuming a gearing of debt:equity of 7:30. The cash cover, interest cover and principal cover ratios are all of reasonably high magnitude and suggest that the project is secure in its ability to service debt. The loan life and project life ratios are well above the recommended levels.

WA_Ni_Sulphide	Year	1	2	3	4	5	6	7	8	9	10
Project Finance											
Annual Cash Flow (M$)		−8.84	−292.14	−193.60	213.69	289.91	358.32	358.30	356.16	354.01	351.87
Loan Charges											
Up Front Fee (M$)		−4.20									
Commitment Fee (% Unused Loan) (M$)		−4.14	−1.74	0.00	0.00	0.00	0.00	0.00	0.00	0.00	0.00
Fixed Loan Charges (M$)		−0.50									
Total Loan Charges		−8.84	−1.74	0.00	0.00	0.00	0.00	0.00	0.00	0.00	0.00
Opening (M$)		6.61	246.08	420.15	420.15	420.15	420.15	336.12	252.09	168.06	84.03
Drawdown (M$)	420.15	6.61	239.47	174.07	0.00	0.00	0.00	0.00	0.00	0.00	0.00
Repayment (M$)		0.00	0.00	0.00	0.00	0.00	−84.03	−84.03	−84.03	−84.03	−84.03
Closing (M$)		6.61	246.08	420.15	420.15	420.15	336.12	252.09	168.06	84.03	0.00
Interest (M$)		−0.56	−20.92	−35.71	−35.71	−35.71	−32.14	−25.00	−17.86	−10.71	−3.57
Debt Service (Interest + Repayment) (M$)		−0.56	−20.92	−35.71	−35.71	−35.71	−116.17	−109.03	−101.89	−94.74	−87.60
Equity Funding Required (M$)	180.07	2.83	102.63	74.60	0.00	0.00	0.00	0.00	0.00	0.00	0.00
Equity Raised (M$)		2.83	105.46	180.07	180.07	180.07	180.07	180.07	180.07	180.07	180.07

(Continued)

Table 10.3. *(Continued)*

WA_Ni_Sulphide	Year	1	2	3	4	5	6	7	8	9	10
Annual Cash Flow After Funding (M$)		0.05	29.04	19.36	177.97	254.20	242.14	249.27	254.27	259.27	264.27
Accumulated Cashflow (M$)		0.05	29.09	48.45	226.42	480.62	722.77	972.04	1226.30	1485.57	1749.84
Debt/Equity Ratio	233.33%										
NPV(Loan Life)	M$				1226.71	1153.01	996.82	839.40	661.48	463.44	243.57
NPV(Project Life)	M$				2092.24	2092.11	2015.74	1944.93	1860.98	1764.90	1655.65
LLR (Loan Life Ratio)	>1.5				2.92	2.74	2.37	2.50	2.62	2.76	2.90
PLR (Project Life Ratio)	>2.0				4.98	4.98	4.80	5.79	7.38	10.50	19.70
Cash Cover (Debt Service Coverage)	>1.5				5.98	8.12	3.08	3.29	3.50	3.74	4.02
Reserves Tail Ratio (RTR >30%)	46.87%				95.02%	88.38%	80.08%	71.78%	63.47%	55.17%	46.87%
ICR (Interest Cover Ratio)					5.98	8.12	11.15	14.33	19.95	33.04	98.53
PCR (Principal Cover Ratio)							3.88	3.97	4.03	4.09	4.14
RC (Residual Cover)							0.79	1.51	2.17	2.78	3.34
Weighted Average Cost of Capital	8.67%										

Table 10.4. Scenario analysis of nickel sulphide project example. IRR for all levels of gearing remains at 46%.

Debt: equity	Loan interest rate (%)	CAPM (%)	WACC (%)	NPV ($ million)
0:100		12		1,081
60:40	8.0	12	8.160	1,212
70:30	8.5	15	8.665	1,008
80:20	9.6	18	8.920	870

Figure 10.7. Palabora Open Pit showing side wall failure. Reproduced with permission of Palabora Mining Company.

Fines from the waste rock are starting to report at the draw points in the underground block-caving operation resulting in dilution and drop-off in head grades. This is discussed in Section 8.7.3. This type of scenario can reduce the reserve tail and is a good example of the convergence of technical and financial risk.

Sensitivity analysis demonstrates that the project can tolerate Capex and OpCosts of up to 20% above those used in the base case without compromising commercial viability. To be complete the financial model should also incorporate functionality that generates an NPV for the different stakeholders that could include a partner with a free-carry, the promoters of the project and fund managers that might expect a higher return on equity than the 12% that is used in the optimised debt:equity scenario of 60:40.

10.6 Longwall Coal Case History and Role of Infrastructure

The longwall mining scenario considered in Section 6.4 takes into account the relationship between fixed mining costs of $15/tonne, plant performance and differential prices for the different product streams. Running the model using IC-CoalEval demonstrates that the return on an investment of $31 million in a longwall mining system is highly dependent on the utilisation of the equipment. The base case assumes that coal will be cut at a rate of 2.4 million tonnes per annum. The cost form is given in Figure 10.8. This scenario generates an NPV of $35 million and an IRR of 33%.

If production is 20% less than the target 2.4 Mt and only 1.92 Mt is produced then, according to the cost estimation empirical formula given in Section 8.3.2, OpCosts would increase from $15/tonne to $17.15/tonne. The NPV would then drop to $6.71 million and the IRR to 18%.

The fundamental problem with any coal project is that the product is a low-value commodity. The only way an investment can generate adequate returns is to aim for a high level of production well beyond that of the base case used. Not only does this amplify the margin between revenue and OpCosts, economies of scale mean that the cost of producing a tonne of coal decreases with an increase in production.

To put this into an international context, in 2012 Sasol invested $250 million in their Thubelisha underground mine in South Africa that is designed to extract coal using room and pillar mining methods. It has the capacity to produce 10 million tonnes of coal per annum. While it could be argued that Sasol's product is used

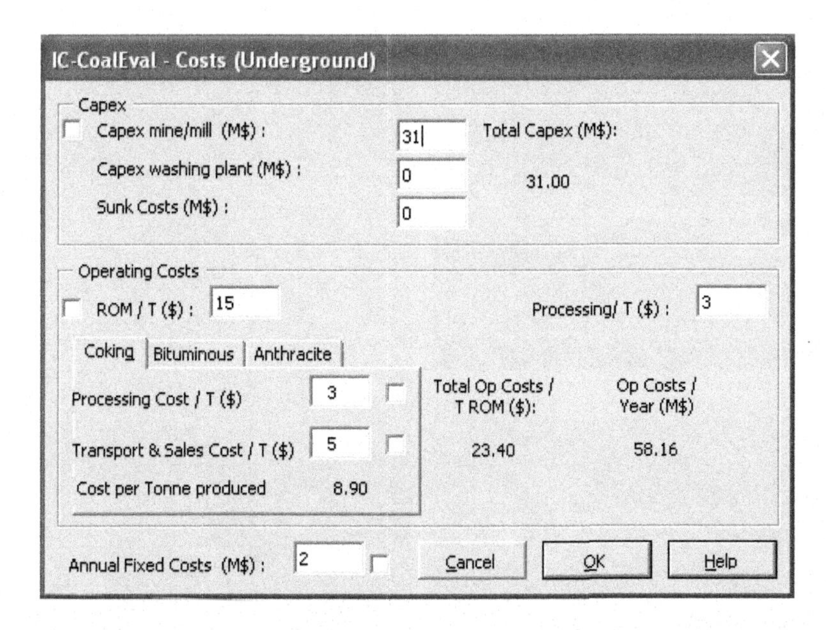

Figure 10.8. Cost menu taken from IC-CoalEval.

in their oil-from-coal operation, AngloAmerican produces coal for the international markets. Their Zibulo Colliery in South Africa produces six million tonnes of coal per annum from underground mining using room and pillar methods. A single section using one continuous miner and two shuttle cars produces 750,000 tonnes of run of mine coal per annum.

A further key factor in the commercial viability of a coal project is transport costs. For a project to derive premium prices associated with an export market requires availability of infrastructure. This would include rail-loading facilities, a dedicated railway and a port such as the Richards Bay facility in South Africa (see Figure 10.9). This level of investment usually needs the involvement of government or the international finance institutions. The similarity with the development of iron ore projects is obvious. The alternative to a major infrastructure development for coal projects is contracts to supply coal for local steam-powered electrical generation. A captive power station would have a quite different relationship to the coal producers than if the latter were captive to the former.

Figure 10.9. Coal infrastructure in South Africa. Top left: loading facility. Bottom left and right: from Transport World Africa website and reproduced with their permission.

10.7 Petroleum Case History

An oil company has received technical data relating to a potential project and now needs to evaluate its viability.

The project relates to a hydrocarbon reservoir containing oil only — no associated gas or condensate occurs. The company's technical experts have undertaken subsurface and engineering studies which suggest it could potentially produce oil for 13 years from the start of production. According to the scenario in the development plan, the production would ramp up from zero at the beginning of year four, to a plateau production rate of 12,000 bpd at the beginning of year five (i.e. a 2-year ramp-up period).

The plateau production would last for 3 years, after which the production rate would decline. The decline can be modelled as a simple 40% decline rate, where the average production rate in any year is 60% of the value in the previous year. The profile will appear as follows in Figure 10.10:

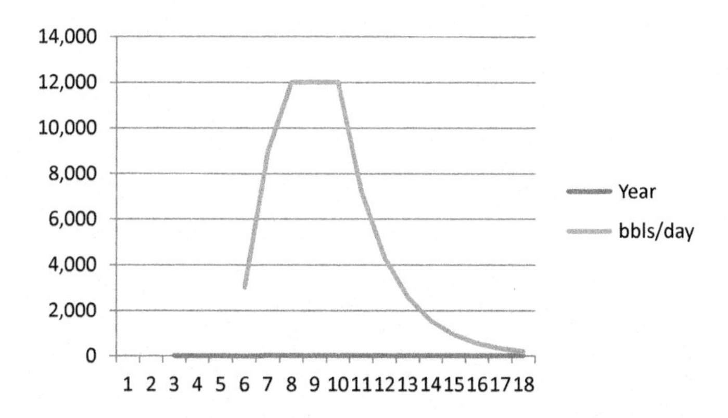

Figure 10.10. Petroleum production profile.

The oil company is the sole "contractor" and is responsible for all OpCosts and Capex. Details of the field development and OpCosts are given below. The government does not invest in the project.

The fiscal regime is a simple tax and royalty concession. There is an oil royalty of 5% of gross revenue. The contractor is also liable for corporate tax at a rate of 20%. The oil royalty is deductible against taxable income. Capex are depreciated using a 5-year straight line method for tax purposes. Tax losses can be carried forward indefinitely and both the royalty and tax liabilities are paid in the year in which they are incurred. Additional information available includes the following:

- Assume the inflation forecast is zero (i.e. do not model inflation).
- Assume a constant oil price of $100 per barrel.
- While the field is producing, it will incur fixed OpCosts of $40 million per year.
- The capital development costs are $50 million in year one, and $500 million in year three.
- There are no decommissioning costs.

- The company uses an "annual end-of-year" discounting method (i.e. assume cash flow in any one year all occurs at the end of each year), and.
- The company's nominal WACC is 6%.

A simple cash flow model for the total project will determine the contractor's and government's cash flows and allows the following metrics to be calculated:

- The field technical oil resource is 23.98 millions of barrels.
- The year of the field economic limit (the final year of economically viable production) is year 12.
- The field economic reserve is 23.24 millions of barrels.
- The total undiscounted project cash flow is $1,413.85 million.
- The NPV of the contractor's net cash flow is $592 million, and.
- The NPV of the government's cash flow (using the contractor's discount rate) is $241 million.

The results of sensitivity analysis on the base case are given in Table 10.5. This demonstrates that while the contractor may benefit more than the government at the higher base case oil price, the opposite happens at the lower price. This may not be an optimal arrangement as government would want a fiscal model that continues to provide incentives to the contractor to continue production during periods of low oil prices.

When we consider petroleum fiscal models the level of complexity in both tax and royalty regimes as well as production-sharing contracts (which are essentially complex tax models) unintended consequences can occur. Agreements then have to be renegotiated. Furthermore, the relatively simple interrelationships between tax given in the financial model for a metal project and the linked economic model breaks down

Table 10.5. Sensitivity analysis of the petroleum model.

Oil price $/bbl	Contractor NPV $ million	Government NPV $ million
100	592	241
75	300	150
50	7	61

in a petroleum fiscal model. Petroleum fiscal models are therefore normally discounted using a flat corporate cost of capital.

10.8 Final Thoughts

Do not introduce complexity in a financial model for its own sake as this can result in a bias that generates an outcome that you want. This is illustrated in Figure 10.11 which is apposite given that as an undergraduate the author did own a Triumph 650 Bonneville motorbike. The key characteristic of a financial model must be utility aimed at supporting robust investment decision-making.

Figure 10.11. Cartoon by Roger Beale first published in the FT's Business Education section on Monday 20 Feb 2012 and reproduced with permission.

Bibliography*

References

Ala, M. (in press). *Introduction to Petroleum Geoscience*, Imperial College Press, London.

Buchanan, D.L. (1988). *Platinum-Group Element Exploration*, Elsevier, Oxford.

Halsall, C. (1989). *Intrusive Magmatism, Volcanism and Massive Sulphide Mineralisation at Rio Tinto, Spain,* PhD dissertation, University of London.

Hayes, C. (2014). *Modern Valuation Techniques Applied to a Development Stage Uranium Junior.* MSc (Metals and Energy Finance) dissertation, Imperial College London.

Houlding, S. (2010). *Practical Geostatistics, Modeling and Spatial* Analysis. Edu-Mine Online Course.

Robb, L. (2004*) Introduction to Ore-Forming Processes*, Wiley-Blackwell, Hoboken, NJ.

Underhill, J.R. and Stoneley, R. (1998). 'Introduction to the Development, Evolution and Petroleum Geology of the Wessex Basin', in Underhill, J.R (ed.), *The Development, Evolution and Petroleum Geology of the Wessex Basin*, Geol. Soc. London, Special Publications, 133, pp. 1–18.

Wells, J. (1998). *Bre-X: The Inside Story of the World's Biggest Mining Scam*, Orion, London.

*As you will note, the bibliography is very short suggesting that I have no need to call on the extensive body of specialist literature that is related to the topics I have addressed. I have, however, attempted to present a personal perspective of the field of metals and energy finance based mainly on my own experience. Although the references given in the bibliography are cited in the text, they are also indicative of the types of published and unpublished sources that supplement the material covered.

Index

Printed in the United States
By Bookmasters